Controlling the Quantum World

The Science of Atoms, Molecules, and Photons

Committee on AMO 2010

Board on Physics and Astronomy

Division on Engineering and Physical Sciences

NATIONAL RESEARCH COUNCIL
OF THE NATIONAL ACADEMIES

THE NATIONAL ACADEMIES PRESS
Washington, D.C.
www.nap.edu

THE NATIONAL ACADEMIES PRESS 500 Fifth Street, N.W. Washington, DC 20001

NOTICE: The project that is the subject of this report was approved by the Governing Board of the National Research Council, whose members are drawn from the councils of the National Academy of Sciences, the National Academy of Engineering, and the Institute of Medicine. The members of the committee responsible for the report were chosen for their special competences and with regard for appropriate balance.

This project was supported by the Department of Energy under Award No. DE-FG02-04ER15610 and by the National Science Foundation under Award No. PHY-0443243. Any opinions, findings, conclusions, or recommendations expressed in this publication are those of the author(s) and do not necessarily reflect the views of the sponsors.

Cover: A purple laser beam slows erbium atoms (the purple beam traveling right to left) emerging from an oven at 1300°C, in preparation for trapping and cooling. SOURCE: National Institute of Standards and Technology.

Library of Congress Cataloging-in-Publication Data

Controlling the quantum world : the science of atoms, molecules, and photons / Committee on AMO 2010, Board on Physics and Astronomy, Division on Engineering and Physical Sciences.
p. cm.
Includes bibliographical references.
ISBN 978-0-309-10270-4 (pbk.)
1. Quantum theory. 2. Atoms. 3. Molecules. 4. Photons. I. National Research Council (U.S.). Committee on Atomic, Molecular, and Optical Sciences 2010.
QC174.12.C67 2006
539—dc22

2007012182

Additional copies of this report are available from The National Academies Press, 500 Fifth Street, N.W., Washington, DC 20001; (800) 624-6242 or (202) 334-3313 (in the Washington metropolitan area); Internet <http://www.nap.edu>; and the Board on Physics and Astronomy, National Research Council, 500 Fifth Street, N.W., Washington, DC 20001; Internet <http://www.national-academies.org/bpa>.

Printed in the United States of America

COMMITTEE ON AMO 2010

PHILIP H. BUCKSBAUM, University of Michigan, *Co-chair*
ROBERT EISENSTEIN, *Co-chair*
GORDON A. BAYM, University of Illinois at Urbana-Champaign
C. LEWIS COCKE, Kansas State University
ERIC A. CORNELL, University of Colorado/JILA
E. NORVAL FORTSON, University of Washington
KEITH HODGSON, Stanford Synchrotron Radiation Laboratory
ANTHONY M. JOHNSON, University of Maryland at Baltimore County
STEVEN KAHN, Stanford Linear Accelerator Center
MARK A. KASEVICH, Stanford University
WOLFGANG KETTERLE, Massachusetts Institute of Technology
KATE KIRBY, Harvard-Smithsonian Center for Astrophysics
PIERRE MEYSTRE, University of Arizona
CHRISTOPHER MONROE, University of Michigan
MARGARET M. MURNANE, University of Colorado/JILA
WILLIAM D. PHILLIPS, National Institute of Standards and Technology
STEPHEN T. PRATT, Argonne National Laboratory
K. BIRGITTA WHALEY, University of California at Berkeley

Consultants to the Committee

NEIL CALDER, Stanford Linear Accelerator Center
NEAL F. LANE, Rice University

Staff

DONALD C. SHAPERO, Director
MICHAEL H. MOLONEY, Study Director
BRIAN D. DEWHURST, Senior Program Associate
PAMELA A. LEWIS, Program Associate
PHILLIP D. LONG, Senior Program Assistant
VAN AN, Financial Associate

Preface

This report is an accounting of the AMO 2010 study undertaken by the National Research Council (NRC) of the National Academies to assess opportunities in atomic, molecular, and optical (AMO) science and technology over roughly the next decade. The charge for this study was devised by a Board on Physics and Astronomy standing committee, the Committee on Atomic, Molecular, and Optical Sciences, in consultation with the study's sponsors, the Department of Energy and the National Science Foundation. The Committee on AMO 2010, which carried out the study, was asked to assess the state of the field of AMO science, emphasizing recent accomplishments and identifying new and compelling scientific questions. The report is a part of the ongoing Physics 2010 decadal survey that is being undertaken by the National Academy's Board on Physics and Astronomy.

The committee that carried out this study and wrote this report is composed of leaders from many different subfields within the AMO physics community, as well as prominent scientists from outside the field. The committee also received valuable advice from consultants Neal Lane, Rice University, and Neil Calder, Stanford Linear Accelerator Center. In addition, the committee received valuable input from the following colleagues: Laura P. Bautz, Nora Berrah, Joshua Bienfang, John Bollinger, Gavin Brennen, Denise Caldwell, John Cary, Michael Casassa, Henry Chapman, Michael Chapman, Charles Clark, Paul Corkum, Philippe Crane, Roman Czujko, Joseph Dehmer, Brian DeMarco, David DeMille, Todd Ditmire, John Doyle, Henry Everitt, Aimee Gibbons, Janos Hajdu, Hashima Hassan, Robert R. Jones, Chan Joshi, William Kruer, Wim Leemans, Anthony Leggett, Steve Leone,

Heather Lewandowski, Jay Lowell, Lute Maleki, Anne Matsuura, Harold Metcalf, Roberta Morris, Gerard Mourou, William Ott, Peter Reynolds, Eric Rohlfing, Steve Rolston, Michael Salamon, Howard Schlossberg, Barry Schneider, David Schultz, Thomas Stoehlker, David Villeneuve, Carl Williams, and Jun Ye.

Significant effort has been made to solicit community input for this study. This was done via town meetings held at the Annual Meeting of the Division of AMO Physics of the American Physical Society (APS) in Lincoln, Nebraska, in May 2005 and the International Quantum Electronics Conference (jointly sponsored by the APS Division of Laser Science, the Optical Society of America, and the Lasers and Electro-optics Society of the Institute of Electrical and Electronics Engineers) in May 2005 in Baltimore, Maryland. The committee also solicited input from the community through a public Web site. The comments supplied by the AMO community through this site and at the town meetings were extremely valuable primary input to the committee.

The federal agencies that fund AMO research in the United States were also solicited for input, through their direct testimony at open meetings and their written responses to requests for information on funding patterns and other statistical data. These data are summarized in Chapter 8 and in the appendixes to the report. Finally, the committee is grateful to the staff at the White House Office of Science and Technology Policy and the Office of Management and Budget, as well as staff from committees of the Congress concerned with funding legislation, who provided important background on connections between AMO science and national science policy.

In November 2005, the NRC released a short interim report from the AMO 2010 Committee, which was intended as a preview of this final document. It summarized the key opportunities in forefront AMO science and in closely related critical technologies, and it discussed some of the broad-scale conclusions of the final report. It also identified how AMO science supports national R&D priorities. The present report reinforces the preliminary conclusions of the interim report and adds a wealth of detail as well as recommendations.

This report reflects the committee's enthusiasm, inspired by the tremendous excitement within the AMO science community about future R&D opportunities. It would not have been written without the extensive and unselfish work of the entire committee, its many consultants, and the NRC staff. We thank them all for their efforts. We particularly wish to thank Michael Moloney for his expertise and dedication and Don Shapero for his experience and wisdom in assisting us to produce this report.

Philip Bucksbaum
Co-chair

Robert Eisenstein
Co-chair

Acknowledgment of Reviewers

This report has been reviewed in draft form by individuals chosen for their diverse perspectives and technical expertise, in accordance with procedures approved by the National Research Council's Report Review Committee. The purpose of this independent review is to provide candid and critical comments that will assist the institution in making its published report as sound as possible and to ensure that the report meets institutional standards for objectivity, evidence, and responsiveness to the study charge. The review comments and draft manuscript remain confidential to protect the integrity of the deliberative process. We wish to thank the following individuals for their review of this report:

Keith Burnett, University of Oxford,
Alexander Dalgarno, Harvard-Smithsonian Center for Astrophysics,
David P. DeMille, Yale University,
Chris H. Greene, University of Colorado,
William Happer, Princeton University,
Wendell T. Hill III, University of Maryland,
Tin-Lun Ho, Ohio State University,
Gerard J. Milburn, University of Queensland,
Richart E. Slusher, Lucent Technologies, and
David J. Wineland, National Institute of Standards and Technology.

Although the reviewers listed above have provided many constructive comments and suggestions, they were not asked to endorse the conclusions or recommendations, nor did they see the final draft of the report before its release. The review of this report was overseen by Daniel Kleppner, Massachusetts Institute of Technology. Appointed by the National Research Council, he was responsible for making certain that an independent examination of this report was carried out in accordance with institutional procedures and that all review comments were carefully considered. Responsibility for the final content of this report rests entirely with the authoring committee and the institution.

Contents

Summary

Atomic, molecular, and optical (AMO) science demonstrates powerfully the ties of fundamental physics to society. Its very name reflects three of 20th century physics' greatest advances: the establishment of the atom as a building block of matter; the development of quantum mechanics, which made it possible to understand the inner workings of atoms and molecules; and the invention of the laser. Navigation by the stars gave way to navigation by clocks, which in turn has given way to today's navigation by atomic clocks. Laser surgery has replaced the knife for the most delicate operations. Our nation's defense depends on rapid deployment using global positioning satellites, laser-guided weapons, and secure communication, all derived directly from fundamental advances in AMO science. Homeland security relies on a multitude of screening technologies based on AMO research to detect toxins in the air and hidden weapons in luggage or on persons, to name a few. New drugs are now designed with the aid of x-ray scattering to determine their structure at the molecular level using AMO-based precision measurement techniques. And the global economy depends critically on high-speed telecommunication by laser light sent over thin optical fibers encircling the globe.[1] These advances, made possible by the scientists in this field, touched many areas of societal importance in the past century, and AMO scientists have been rewarded with numerous Nobel prizes over the past decade, including the 2005 prize in physics.

[1]For further detail on the connections between AMO science and society's needs, see National Research Council, *Atoms, Molecules, Light: AMO Science Enabling the Future*, Washington, D.C.: The National Academies Press (2002), available at <http://www.nap.edu/catalog/10516.html>.

The purpose of this report is to identify the most promising future opportunities in AMO science based on what is known at this time. Building on these findings, the report describes the most fertile avenues for the next decade's research in this field.

Despite a century of phenomenal progress in science, the universe is still a mysterious place. Many fundamental questions remain. One of the most important is that the fundamental forces of nature that shape the universe are still not fully understood. New AMO technology will help provide answers in the coming decades—in precision laboratory measurements on the properties of atoms, in giant gravitational observatories on Earth, or in even larger observatories based in space. Tremendous advances in precision timekeeping also place us at the threshold of answering some of the central questions.

Society has other urgent needs that AMO physics is poised to address. How will we meet our energy needs as Earth's natural resources become depleted and the environment changes? Solar energy collection and conversion, laser fusion, or molecular biophysics may offer solutions, and all of these have strong connections to AMO science. Health threats are likely to increase on our interconnected and highly populated planet, and rapid response to new contagions requires the development of ways to detect biomolecules remotely, possibly through advanced laser techniques, as well as ways to measure their structure and chemistry, a priority effort at advanced x-ray light sources. The future security of our nation's most powerful weapons may depend on our ability to reproduce the plasma conditions of a fusion bomb in the tiny focus of a powerful laser. And, controlling that plasma is key to harnessing its power for beneficial uses.

These last lines underscore how AMO science contributes strongly to the development of advanced technologies and tools. Instruments made possible by AMO science and related technical developments are today everywhere in experimental science—from astronomy to zoology. In many instances they enable revolutionary experiments or observations that lead to revolutionary new insights. A century of progress toward understanding the mysterious and counterintuitive nature of quantum mechanics now places AMO science at the vanguard of a new kind of quantum revolution, in which *coherence* and *control* are the watchwords.

SIX COMPELLING RESEARCH OPPORTUNITIES FOR AMO SCIENCE

This report concludes that research in AMO science and technology is thriving. It identifies, from among the many important and relevant issues in AMO science, six broad grand challenges that succinctly describe key scientific opportunities available to AMO science:

- Revolutionary new methods to measure the nature of space and time with extremely high precision have emerged within the last decade from a convergence of technologies in the control of the coherence of ultrafast lasers and ultracold atoms. This new capability creates unprecedented new research opportunities.
- Ultracold AMO physics was the most spectacularly successful new AMO research area of the past decade and led to the development of coherent quantum gases. This new field is poised to make major contributions to resolving important fundamental problems in condensed matter science and in plasma physics, bringing with it new interdisciplinary opportunities.
- High-intensity and short-wavelength sources such as new x-ray free-electron lasers promise significant advances in AMO science, condensed matter physics and materials research, chemistry, medicine, and defense-related science.
- Ultrafast quantum control will unveil the internal motion of atoms within molecules, and of electrons within atoms, to a degree thought impossible only a decade ago. This is sparking a revolution in the imaging and coherent control of quantum processes and will be among the most fruitful new areas of AMO science in the next 10 years.
- Quantum engineering on the nanoscale of tens to hundreds of atomic diameters has led to new opportunities for atom-by-atom control of quantum structures using the techniques of AMO science. There are compelling opportunities in both molecular science and photon science that are expected to have far-reaching societal applications.
- Quantum information is a rapidly growing research area in AMO science and one that faces special challenges owing to its potential application in data security and encryption. Multiple approaches to quantum computing and communication are likely to be fruitful in the coming decade, and open international exchange of people and information is critical in order to realize the maximum benefit.

Surmounting these challenges will require important advances in both experiment and theory. Each of these science opportunities is linked closely to the new tools that will also help in meeting critical national needs. The key future opportunities for AMO science presented by these six grand challenges are based on the rapid and astounding developments in the field, a result of investments made by the federal R&D agencies in AMO research programs. These compelling grand challenges in AMO research are discussed in more detail in the report, which also highlights the broad impact of AMO science and its strong connections to other branches of science and technology and discusses the strong coupling to national priorities in health care, economic development, the environment, national defense, and homeland security. Finally, the report analyzes trends in federal support for research, compiled from responses provided by AMO program officers at federal agencies.

The linkages between opportunities for AMO science and technology and national R&D goals are clear. The White House set forth the country's R&D priorities in the July 8, 2005, memorandum written by the science advisor to the President and the director of the Office of Management and Budget. These priorities were reiterated and strengthened in the President's State of the Union Address in January 2006 and in the President's Budget Request for FY2007. AMO scientists contribute to these national priorities in several key areas:

- Advancing fundamental scientific discovery to improve the quality of life.
- Providing critical knowledge and tools to address national security and homeland defense issues and to achieve and maintain energy independence.
- Enabling technological innovations that spur economic competitiveness and job growth.
- Contributing to the development of therapies and diagnostic systems that enhance the health of the nation's people.
- Educating in science, mathematics, and engineering to ensure a scientifically literate population and qualified technical personnel who can meet national needs.
- Enhancing our ability to understand and respond to global environmental issues.
- Participating in international partnerships that foster the advancement of scientific frontiers and accelerate the progress of science across borders.
- Contributing to the mission goals of federal agencies.

In discussing the state of AMO science and its relation to the federal government, the report offers some observations and conclusions. Given the budget and programmatic constraints, generally the federal agencies questioned in this study have managed the research profile of their programs well in response to the opportunities in AMO science. In doing so, the agencies have developed a combination of modalities (large groups, centers and facilities, and expanded single-investigator programs). Much of the funding increase that has taken place at the Department of Energy (DOE), the National Institute of Standards and Technology (NIST), and the National Science Foundation (NSF) has served to benefit activities at research centers. The overall balance of the modalities for support of the field has led to outstanding scientific payoffs. In addition, the breadth of AMO science and the range of the agencies that support it are exceedingly important to future progress in the field and have been a key factor in its success so far.

On the other hand, the committee notes with concern the decline in research funding in general and in basic research funding in particular (the so-called 6.1 budget) at Department of Defense (DOD) agencies. This is troubling especially because fundamental scientific research has been a critical part of the nation's defense strategy for more than half a century.

Since all of the agencies questioned by the committee reported that they receive substantially more proposals of excellent quality than they are able to fund, it appears that AMO science remains rich with promise for future progress. The committee concludes that AMO science will continue to make exceptional advancements in science and in technology for many years to come.

A substantial increase in the nation's investment in the physical sciences has been identified as a national priority with vast importance for national security, economic strength, health care, and defense.[2] As the President has indicated, a program of increased investment must be directed at both improving education in the physical sciences and mathematics at all levels as well as significantly strengthening the research effort. Such a program will enhance the nation's ability to capture the benefits of AMO science. Support for basic research is a vital component of the nation's defense strategy. The recent decline in research funding at the defense-related agencies, most particularly in funding for basic research, is harming the nation. Industry-sponsored basic research also plays a key role in enabling technological development, the committee concludes, and steps should be taken to reinvigorate it.

This report notes three key committee findings in programmatic issues:

[2]See the following reports: House Committee on Science, *Unlocking Our Future: Toward a New National Science Policy* (1998), available at <http://www.house.gov/science/science_policy_report.htm>, accessed June 2006; National Commission on Mathematics and Science Teaching for the 21st Century (Glenn Commission), *Before It's Too Late* (2000), available at <http://www.ed.gov/inits/Math/glenn/report.pdf>, accessed June 2006; United States Commission on National Security/21st Century, *Road Map for National Security: Imperative for Change* (2001) (also known as the Hart-Rudman report), available at <http://www.fas.org/man/docs/nwc/phaseiii.pdf>, accessed June 2006; National Science Foundation, *The Science and Engineering Workforce: Realizing America's Potential* (2003), available at <http://www.nber.org/~sewp/>, accessed June 2006; NAS/NAE/IOM, *Rising Above the Gathering Storm: Energizing and Employing America for a Brighter Economic Future*, Washington, D.C.: The National Academies Press (2007); U.S. Domestic Policy Council, *American Competitiveness Initiative* (2007), available at <http://www.whitehouse.gov/stateoftheunion/2006/aci/aci06-booklet.pdf>, accessed June 2006.

- The extremely rapid increase in technical capabilities and the associated increase in the cost of scientific instrumentation have led to very significant added pressures (over and above the usual Consumer Price Index inflationary pressures) on research group budgets. In addition, not only has the cost of instrumentation increased, but also the complexity and challenge of the science make investigation much more expensive. This "science inflator" effect means that while it is now possible to imagine research that was unimaginable in the past, finding the resources to pursue that research is becoming increasingly difficult.
- In any scientific field where progress is extremely rapid, it is important not to lose sight of the essential role played by theoretical research. Programs at the federal agencies that support AMO theory have been and remain of critical importance. NSF plays a critical and leading role in this area, but its support of AMO theoretical physics is insufficient.
- AMO science is an enabling component of astrophysics and plasma physics but is not adequately supported by the funding agencies charged with responsibility for those areas.

The committee made a number of findings on workforce issues. It agrees with many other observers that the number of American students choosing physical sciences as a career is dangerously low. Without remediation, this problem is likely to open up an unacceptable expertise gap between the United States and other countries. Since AMO science offers students an opportunity for exceptionally broad training in a field of great importance, and therefore of excellent job prospects, it is poised to contribute to a solution of the problem. The committee points out that any effort to attract more American-born students into the physical sciences must recognize that personnel adjustments occur on a timescale of decades. Reversing the decline will require a long-term effort.

It must be remembered, too, that it will always be in the national interest to attract and retain foreign students in the physical sciences. Similarly, the report notes that scientists and students in the United States derive great benefits from close contact with the scientists and students of other nations that takes the form of international collaborations, exchange visits, meetings, and conferences. These activities are invaluable for promoting both excellent science and better international understanding, and they support the economic, educational, and national security needs of the United States. It is, therefore, essential to U.S. interests that these activities continue.

RECOMMENDATIONS

Finally, the committee offers six recommendations that form a strategy to realize fully the potential at the frontiers of AMO science:

Recommendation. In view of the critical importance of the physical sciences to national economic strength, health care, defense, and domestic security, the federal government should embark on a substantially increased investment program to improve education in the physical sciences and mathematics at all levels and to strengthen significantly the research effort.

Recommendation. AMO science will continue to make exceptional contributions to many areas of science and technology. The federal government should therefore support programs in AMO science across disciplinary boundaries and through a multiplicity of agencies.

Recommendation. Basic research is a vital component of the nation's defense strategy. The Department of Defense, therefore, should reverse recent declines in support for 6.1 research at its agencies.

Recommendation. The extremely rapid increase in the technical capability of scientific instrumentation and its cost has significantly increased pressures (over and above the usual Consumer Price Index inflationary pressures) on research budgets. The federal government should recognize this fact and plan budgets accordingly.

Recommendation. Given the critical role of theoretical research in AMO science, the funding agencies should reexamine their portfolios in this area to ensure that the effort is at proper strength in workforce and funding levels.

Recommendation. The federal government should implement incentives to encourage more U.S. students, especially women and minorities, to study the physical sciences and take up careers in the field. It should continue to attract foreign students to study physical sciences and strongly encourage them to pursue their scientific careers in the United States.

1

Controlling the Quantum World: AMO Science in the Coming Decade

Atomic, molecular, and optical (AMO) science demonstrates powerfully the ties of fundamental physics to society. Its very name reflects three of 20th century physics' greatest advances: the establishment of the atom as a building block of matter; the development of quantum mechanics, which made it possible to understand the inner workings of atoms and molecules; and the invention of the laser, which changed everything from the way we think about light to the way we store and communicate information. The field encompasses the study of atoms, molecules, and light, including the discovery of related applications and techniques. This report illustrates how AMO science and technology touches almost every sphere of societal importance—navigation using the latest atomic clocks; surgery with a host of new laser tools; ensuring the nation's defense using global positioning satellites and secure communication; defending the homeland with screening technologies to detect toxins in the air and hidden weapons in luggage or on persons; improving health care with improved drug design tools and new diagnostic scanners; and underpinning the world's economies with a global communications network based on high-speed telecommunication by laser light.[1]

The immense advances in science over the past century have only just begun to explain the mysteries of the universe. One of the primary goals of AMO science is to

[1]For further detail on the connections between AMO science and society's needs, see National Research Council, *Atoms, Molecules, Light: AMO Science Enabling the Future*, Washington, D.C.: The National Academies Press (2002), available at <http://www.nap.edu/catalog/10516.html>, accessed June 2006.

reveal the workings of nature on a fundamental level. In addition, society continues to have many urgent challenges that AMO research seeks to address. The unifying thread between the pure and applied work is quantum mechanics: AMO research develops tools and seeks knowledge on the quantum level, enabling progress in many other fields of science, engineering, and medicine.

The overarching emerging theme in AMO science is control of the quantum world. The six broad grand challenges outlined in this report describe key scientific opportunities in the coming decade. They are precision measurements; ultracold matter; ultra-high-intensity and short-wavelength lasers; ultrafast control; nanophotonics; and quantum information science. These challenges will drive important advances in both experiment and theory. Each of these science opportunities is linked closely to new tools that will also help in meeting critical national needs (see Figure 1-1).

WHAT IS THE NATURE OF PHYSICAL LAW?

What are the undiscovered laws of physics that lie beyond our current understanding of the physical world? What is the nature of space, time, matter, and energy? AMO science provides exquisitely sensitive tools to probe these questions. For example, a force that alters the fundamental forward-backward symmetry of time has been studied extensively by high energy physicists, but another such force beyond the current Standard Model of the universe is now widely expected to exist. This tiny but revolutionary effect could show up first in the next decade in AMO experiments that look for deviations in the nearly perfect spatial symmetry found in atoms. A second question asks whether the laws of physics are constant over time or across the universe. A new generation of ultraprecise clocks will enable laboratory searches for time variations of the fundamental constants of nature. Answers will also come from AMO research that is helping to interpret astrophysical observations of the most exotic and most distant realms in the universe. The advanced technologies developed for such fundamental physics experiments have many other uses. They will improve the accuracy of direct gravity-wave detection and of next-generation global positioning satellites and will produce new medical diagnostics. These advances are described briefly in the next paragraphs and explored more fully in Chapter 2.

Since the atomic concept was finally accepted at the beginning of the 20th century, atoms have proven central to the discovery and understanding of the laws of physics. Today remarkably sensitive techniques probe the properties of atoms, molecules, and light over enormous ranges: from submicroscopic to cosmic distances, in both familiar environments and the most exotic realms in the universe. The unprecedented sensitivity with which these fundamental properties can be

FIGURE 1-1 NIST-F1 is a cesium fountain atomic clock in Boulder, Colorado. Along with other international atomic clocks, NIST-F1 helps define the official world time standard. In 2000 the uncertainty in the clock's accuracy was about 1×10^{-15}, but by the summer of 2005, the uncertainty was reduced to about 5×10^{-16}, which means it would not gain or lose more than a second in more than 60 million years. For more information on this clock, see <http://www.tf.nist.gov/cesium/fountain.htm>. SOURCE: National Institute of Standards and Technology, copyright Geoffrey Wheeler.

measured is not only advancing science but is also yielding new technology for applications as diverse as studying the brain and detecting lung disease, for terrestrial guidance and space navigation, and for mapping local gravitational fields and detecting subsurface features in the Earth.

AMO experiments could provide an understanding of a fundamental property of time and physical law. How the laws of physics might change if time went backward is not just a whimsical question from science fiction; it is one of the most vigorously debated questions in the physics of fundamental forces. The measurement of atomic electric dipole moments (EDMs) could provide an answer to this question. The EDM is a tiny separation between the centers of positive and negative charges in an atom, which has been predicted by nearly every class of advanced theory in particle physics, including supersymmetry. EDMs have never been observed and must be very tiny if they even exist; we do, however, possess the technology that could allow their detection in the next decade. They would reveal new physics beyond our current understanding of the subatomic nature of our universe, as described by the so-called Standard Model. While much of our knowledge about the Standard Model of fundamental interactions comes from high-energy particle accelerators, AMO experiments have provided critical complementary information.

Unprecedented precision has practical consequences. The techniques developed for these fundamental experiments are now surpassing low-temperature superconducting quantum interference devices (SQUIDs) in the precise measurement of magnetic fields, reaching sensitivities better than 10 parts per trillion of Earth's magnetic field. Such sensitivity will improve our ability to measure more accurately the weak magnetic fields of the brain and the heart, thereby helping to diagnose epilepsy, cardiac arrhythmias, and other diseases. Similarly, advances in measuring the magnetic properties of the atoms of noble gases are opening up a new field in medical imaging that will allow high-resolution studies of the lung. Such images cannot be obtained using standard MRI techniques. Current devices based on this new diagnostic tool promise enormous improvement in the early diagnosis of lung disease.

Extraordinary advances in optical spectroscopy are leading to superb atomic clocks. Ultrashort pulsed laser sources have been exploited to create an "optical comb" spanning the entire visible and near-infrared spectrum. With this revolutionary development (recognized by the Nobel prize in 2005) it is possible to count optical frequencies (about 10^{15} Hz) literally in cycles per second and to measure the ratio of optical frequencies with unprecedented precision. New ultra-accurate clocks will test whether fundamental "constants" of nature are changing over time. They also have many direct and near-term technological impacts, including enhancement of the performance of high-end analog-to-digital converters in advanced radar, more accurate global positioning satellites, and many other applications.

Optical and atom interferometry will lead to new navigation tools and measurements of gravitation. New AMO devices are enabling ever more precise measurements of motion by detecting tiny changes in the interference not just between beams of light but also between beams of atoms, as discussed below. Interferometers are the cornerstone of gravitational wave observatories on Earth and in space that are expected to provide new insight into the structure of our universe. Ring laser and fiber-optic gyroscopes are now standard sensors that play a broad role in state-of-the-art navigation systems. Matter-wave interferometers promise a huge improvement in navigational systems accuracy. Laser-based gravimeters are being used worldwide to characterize Earth's gravitational field for the management of oil deposits and other resources. Future systems based on atom-wave interference will enable airborne characterization of gravitational anomalies at unprecedented levels to detect hostile underground structures and tunnels.

Atomic data and atomic theory provide critical support in astrophysics exploration. Our universe serves as an extraterrestrial laboratory in which to test the laws of physics under extreme conditions. Satellite observatories can probe the environments near black holes and the surfaces of neutron stars. Studying the universe can provide clues to the nature of fundamental physical laws at times and at energies than cannot be reached with today's earthbound laboratory experiments. AMO science plays a central role in helping us to understand what the data from radio, optical, and x-ray telescopes are telling us about these extreme astrophysical environments. Collisions of atoms, molecules, electrons, and ions in these extreme regimes yield new spectral features that can be modeled by theorists and used to understand the full range of extraordinary conditions observed in the universe.

As the following chapters show, there are many areas in which AMO transcends disciplinary lines and provides techniques and data that improve both our understanding of the universe and our daily lives. For example, AMO data lie at the heart of the development of plasma processing, efficient lighting, and many other high-temperature chemical reactions. Exciting new developments in biology include results from electron-molecule scattering, where it has been recently discovered that resonant dissociative electron capture plays an important role in radiation damage through DNA strand breaking. Some of these kinds of cross-cutting possibilities are discussed in summary form in the NRC report *Atoms, Molecules, and Light: AMO Science Enabling the Future.*[2]

[2]For further detail on the connections between AMO science and society's needs, see National Research Council, *Atoms, Molecules, and Light: AMO Science Enabling the Future*, Washington, D.C.: The National Academies Press (2002), available at <http://www.nap.edu/catalog/10516.html>.

WHAT HAPPENS AT THE LOWEST TEMPERATURES IN THE UNIVERSE?

The Bose-Einstein condensates (BECs) developed in the laboratories of AMO physicists in the last decade are the coldest objects that have ever existed anywhere in the universe. These remarkable clouds of trapped atoms are about a billionth of a degree above absolute zero, much colder than the dark, frigid, furthest reaches of intergalactic space. Furthermore, BECs and their close cousins, ultracold degenerate Fermi-Dirac gases, are not just cold; these quantum condensates are proving to be very special states of matter (see Figure 1-2). Scientists have discovered that

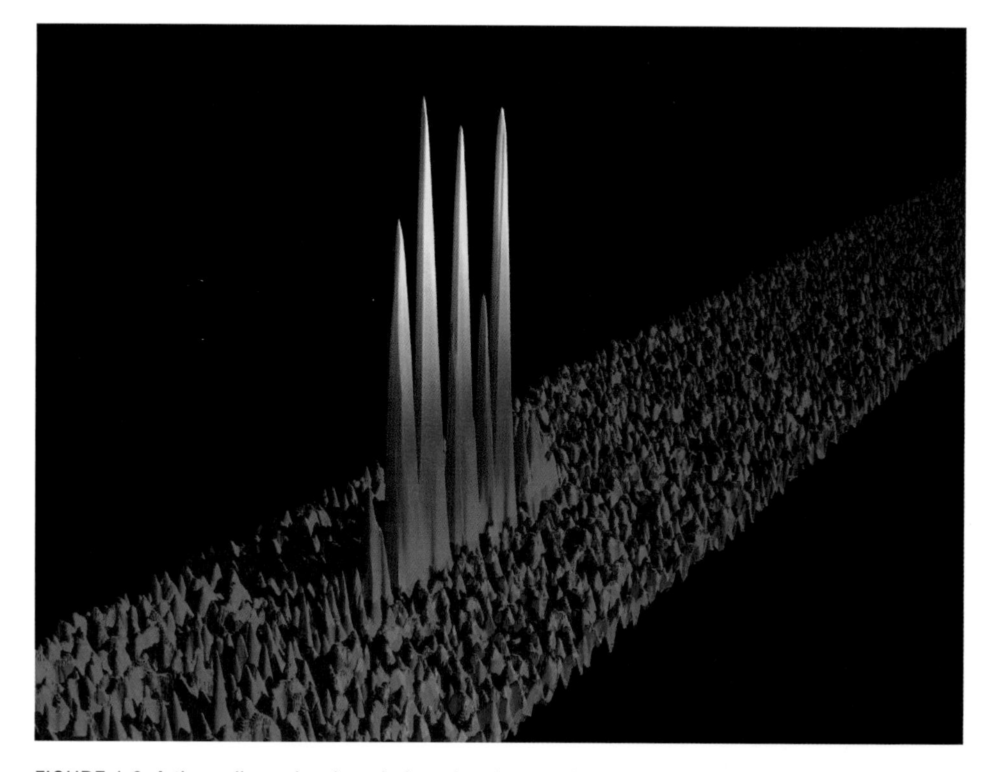

FIGURE 1-2 A three-dimensional rendering of an image of a matter-wave soliton train. Each peak in the train is a Bose-Einstein condensate of lithium atoms, held together by the atoms' self-attraction. The spikes are called matter-wave solitons. Solitons are localized bundles of waves constrained to move in only one dimension and that propagate without spreading. These atomic lithium solitons occur because the ultracold atoms bunch together due to the quantum forces they exert on each other. SOURCE: K.E. Strecker and R.G. Hulet, Rice University.

they have strange and wonderful properties, and in the next decade we can expect a rich harvest of interesting new physics ideas and applications—from technological breakthroughs such as clocks and inertial sensors of unprecedented accuracy, to insights into the physics of ordinary matter as well as matter under extreme conditions. More information on cold quantum gases is contained in Chapter 3 and summarized below.

When breakthrough science happens, it defines a new frontier. Today, AMO science is camped on one of the most exotic frontiers in science—the push toward ever lower temperatures obtained in atomic physics labs. In the last decade, six physicists have won the Nobel prize for their work at the frontier of ultracold atomic gases. The record low temperature stands, as of early 2006, at about a billionth of a degree above absolute zero. By contrast, intergalactic space is a relatively hot 2.7 degrees above absolute zero owing to the existence of the cosmic microwave background.

An ultra-low-temperature gas is a fruitful frontier to explore for two reasons: The atoms and molecules are nearly free from thermal fluctuations, and the quantum (or de Broglie) wavelength of the particles becomes extremely large. As a result, the field of ultracold atoms has become a remarkable meeting place for scientists of many different professional specialties, all of whom have come to realize how much can be learned from this new discipline, and all of whom bring their own expertise to the mix.

When a gas of atoms is cooled to "degeneracy"—that is, it becomes so cold that the quantum wave of one atom starts to overlap with that of its nearest neighbor—and the atoms in question are bosons, the result is a BEC. In such a gas, all of a macroscopic fraction of the particles can simultaneously occupy the lowest-energy "wave" in the box in which the atoms are kept.

Ultracold gases offer the intriguing possibility of building completely controllable models for matter. They can be confined in a variety of geometries, including situations that are effectively one- or two-dimensional rather than three-dimensional. It is also possible to confine them in ways that mimic the periodic structure in solid crystals. These uses of ultracold atoms as quantum simulators are among the most exciting recent developments in AMO science. They have a considerable impact on precision measurements, as discussed in Chapter 2, and on quantum information science, as discussed in Chapter 7. Also, these applications occur at the intersection of AMO physics and condensed matter and indeed AMO physics and other fields of physics. These strengthening links between AMO and other parts of physics promise exciting new discoveries in AMO science in particular and in physics more broadly.

WHAT HAPPENS AT THE HIGHEST TEMPERATURES IN THE UNIVERSE?

Lasers in the next decade will reach peak powers of a million billion watts concentrated in a single beam of light for a little more than a millionth of a billionth of a second. This power exceeds, for an instant, all the electrical power production on Earth. The huge electric fields in these focused beams overwhelm the forces that bind electrons in atoms and molecules, leading to exotic states of matter usually found only in neutron stars, the early universe, hydrogen bombs, and particle accelerators. These lasers will help unravel the violent forces we see in the universe around us. High-powered optical lasers have applications to many other important technological problems as well, ranging from the prospect of controlling nuclear fusion as a source of clean, abundant energy to creation of next-generation compact x-ray microscopes with unprecedented resolution.

Advanced lasers open new scientific and technological frontiers that benefit from two closely related technical advances: the development of optical-wavelength laser beams with unprecedented power and a new generation of lasers that produce coherent x rays. Both high-powered lasers and x-ray lasers will expand our knowledge as they extend our use of electromagnetic radiation.

Bright, ultrafast sources of x rays will revolutionize the study of matter in the next decade. Synchrotron x-ray light sources have been important tools for determining structure at the atomic scale. Today there are dozens of accelerator storage rings around the world, heavily used and devoted to research in materials science and chemistry as well as biology and medicine. Another revolutionary tool will become available in the next decade: the x-ray free-electron laser. These x-ray lasers will be more than one billion times brighter than the brightest synchrotrons, with pulses more than one thousand times shorter. This means that they will be capable of concentrating unprecedented energy on the atomic scale in chemicals, materials, and biological molecules. An important challenge of the next decade is to find ways to take full advantage of this new capability to advance chemistry, biology, and medicine. One particularly important application, biomolecular imaging, is discussed in Chapter 5. Other applications are covered in Chapter 4.

New scientific and technological frontiers will be explored using ultraintense visible lasers, that can reach peak powers in excess of 1 million billion watts. Focused beams from the highest-powered lasers can concentrate the equivalent power of the entire electrical grid of the United States onto a spot only a tenth of a human hair in diameter. The enormous electric fields in these focused laser beams dwarf the forces that bind electrons in atoms and molecules, literally tearing them apart in an instant. Such energetic states of matter are usually found only in the most exotic places in the universe—in the center of stars or in the explosion of a nuclear weapon. Scientists are learning how to harness the electric fields generated by intense lasers to create directed beams of electrons, positrons, or neutrons for medical and materials diagnostics.

New ultraintense laser sources can accelerate electrons to high energies in shorter distances than any other method yet devised, opening up the possibility of building powerful particle accelerators in quite small spaces (see Figure 1-3). The accelerated beams have been employed to make radioisotopes for medical use. These lasers have also generated plasmas capable of nuclear fusion, and their light has been converted to x rays for research on dynamics in laser-excited solids. These applications are discussed in more detail in Chapter 4. Any one of these demonstration experiments could become the basis for expanded research and technology in the coming decade.

Controlled thermonuclear fusion is one particularly important challenge from the past decade that will continue to be important in the next. Laser-heated plasmas

FIGURE 1-3 A plasma channel shown in blue and white, about a millimeter long and denser toward the edges, guides the laser and allows it to form high-quality electron beams. As the laser pulse travels from left to right, it traps and accelerates bunches of electrons to high energies in very short distances. SOURCE: W. Leemans, Lawrence Berkeley National Laboratory.

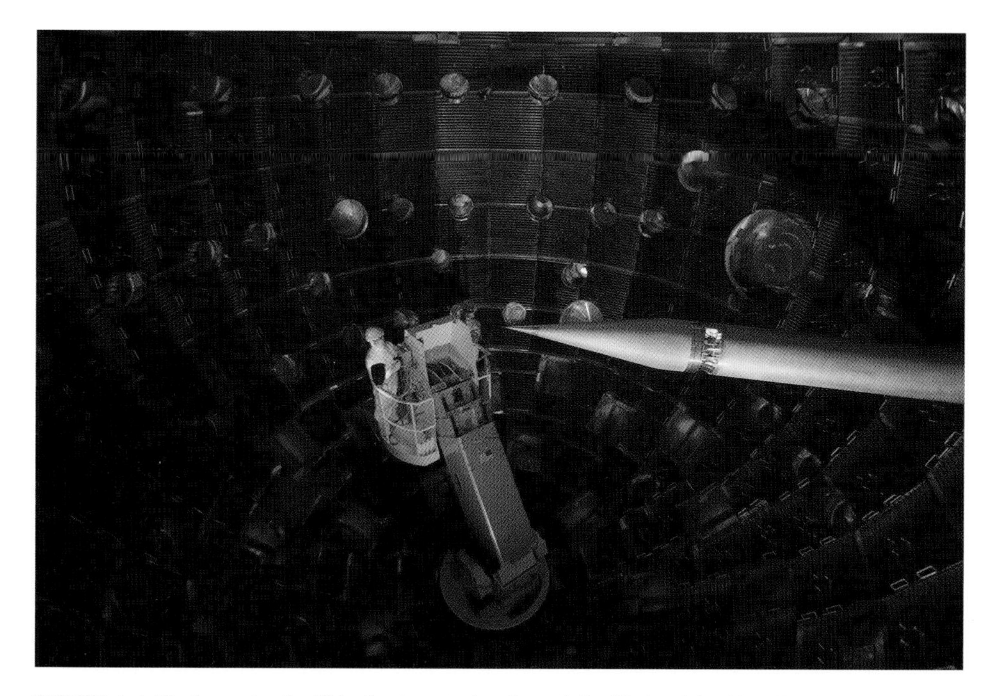

FIGURE 1-4 Workers stand within the target chamber of the National Ignition Facility (NIF), which measures 30 feet in diameter and will contain 192 high-power laser beams, all directed at one small target. NIF can thus deposit enormous amounts of energy in very small volumes over very short times. SOURCE: University of California, Lawrence Livermore National Laboratory.

for fusion energy have been under development in the United States and Europe for decades, but recent progress in ultrafast and high-field lasers holds particular promise for rapid advances toward a device that will produce more fusion energy than it consumes in heating the plasma (the so-called breakeven point). Advanced high-intensity lasers may be a key technology to achieve breakeven. This will be tested in the next decade (see Figure 1-4).

CAN WE CONTROL THE INNER WORKINGS OF A MOLECULE?

In the next decade we will begin to observe the processes of nature as they play out over times shorter than a millionth of a billionth of a second (less than 1 femtosecond—that is, in the attosecond regime). This remarkable new capability is enabled by advances in ultrafast laser- and accelerator-based x-ray strobes, which can detect the motion of electrons in atoms and molecules. Scientists anticipate the possibility of capturing images of motion inside a molecule or taking a snapshot of a protein or virus (see Figure 1-5). We will also be able to control physical phenomena on all of the timescales relevant to atomic and molecular physics, chemistry, biology, and materials science. These previously unavailable tools of quantum control could help tailor new molecules for applications in health care, energy, and security.

A frontier of AMO science is to observe the basic processes of chemistry and biology on the scale of a single molecule. A key ingredient in taking slow-motion pictures is the ability to freeze the action by recording images with a shutter speed much faster than the motion of the object of interest. In atomic and molecular motion, as elsewhere in high-speed photography, the mechanical shutter has been replaced by a short pulse of light, which acts as a stroboscope. The picosecond (or faster) processes within molecules require very short pulses that can only be produced by a laser. The rotational motion of the molecules in a gas cell can be captured by illuminating the gas with laser pulses that are a fraction of a picosecond in duration, while freezing the vibrational motion of molecules requires pulses of a few femtoseconds. Freezing the motion of electrons as they move about the molecule requires subfemtosecond, or attosecond, laser pulses.

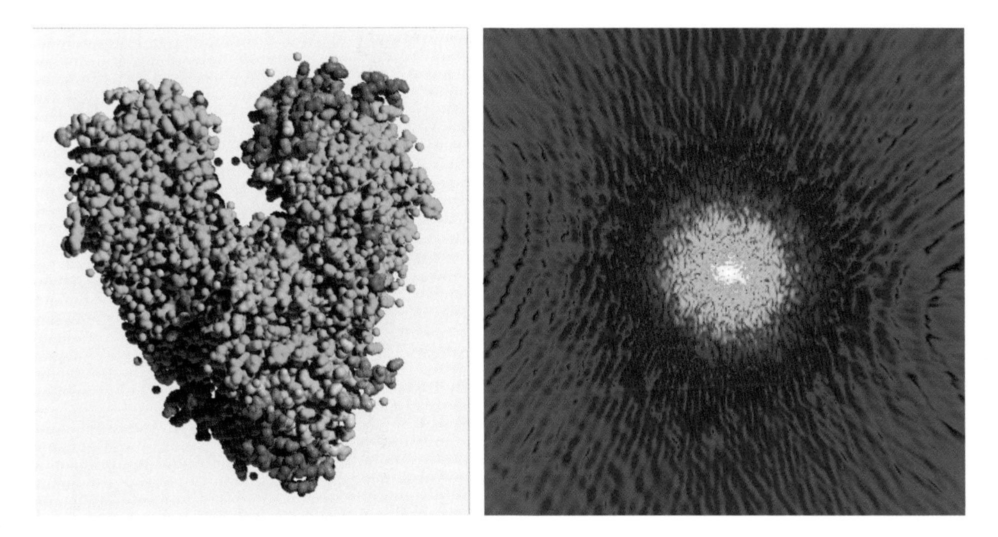

FIGURE 1-5 An x-ray free-electron laser (XFEL) can image single biomolecules. *Left:* a simulated model of the anthrax lethal factor protein. *Right:* A simulation of the diffraction pattern of the molecule on the left, which can be processed by a computer to generate an image of the molecular structure with atomic resolution. SOURCE: Lawrence Livermore National Laboratory.

Capturing the motion of atoms in the attosecond regime is now possible. Rapid molecular vibrations can be observed with a sequence of two subpicosecond pulses in which the first pulse excites the molecule and the second pulse probes the resulting molecular response as a function of the time interval between the pulses. The science breakthroughs brought about by this advance were recognized by the 1999 Nobel prize in chemistry. At the beginning of 2006 strobes as short as one hundred attoseconds have even captured the motion of the electrons within an atom. This motion is the fundamental physical basis for chemistry. Why do some atoms bind, and others not? Why do reactions take the time they do, and why do molecules bend one way but not another? Watching the steps in the dance of electrons will provide a wealth of new insight into the mechanisms of chemistry.

Ultrafast pulses of x rays show great promise for investigating the structure of complex molecules. X rays are very short wavelength light rays with two important differences from ordinary light: They can penetrate ordinary matter and reveal the interiors of solid objects and they can resolve very small objects, down to single atoms. X rays were used to reveal the double helix shape of the DNA molecule, which encodes the genetic information for all living things. Currently the brightest x-ray sources can reveal the atomic details of complex molecules and solids in a static or resting state but are not bright enough to capture changes such as chemical catalysis or absorption of sunlight in photosynthesis.

The new x-ray laser tools of 2010 and beyond will help scientists understand, manipulate, and exploit the molecular universe. What makes the molecules in all living organisms so efficient at carrying out particular tasks? Can we design other molecules to be as efficient as the ones that nature has optimized? What makes certain materials effective catalysts in chemical reactions or gives them the remarkable ability to capture sunlight efficiently and turn it into chemical energy?

New 21st-century tools also place us on the verge of the new discipline of quantum control. This development is enabled by key advances in laser technology, which let us generate light pulses whose shape, intensity, and color can be programmed with unprecedented flexibility. Our ability to control the positions, velocities, and relative spatial orientations of individual atoms and molecules has led to a broad array of precision measurement technologies and devices, leading to a wide range of experiments and discussed throughout this report, that reveal qualitatively new phenomena. A new capability to manipulate the inner workings of molecules is emerging: Lasers can now be used to control the outcome of selected chemical reactions. This control technology may ultimately lead to powerful tools for creating new molecules and materials tailored for applications in health care, nanoscience, environmental science, energy, and national security.

HOW WILL WE CONTROL AND EXPLOIT THE NANOWORLD?

The nanoworld lies in between the familiar classical world of macroscopic objects and the quantum world of atoms and molecules. Nanostructures can have counterintuitive but useful optical properties that arise because they are smaller than the wavelength of light used to observe them. Scientists see unique opportunities to tailor material properties for efficient optical switches, light sources, and photoelectric power generators. Nanomaterials promise the development of single-photon sources and detectors, photonic crystals, environmental sensors, biomedical optics, and novel cancer therapies involving localized optical absorption. These opportunities are described in Chapter 6 and briefly summarized below.

A myriad of powerful new tools are now available to create, visualize, and control structures on the nanoscale. Nanoscience includes fundamental research on the unique phenomena and processes that occur at the nanometer scale (see Figure 1-6). Opportunities that lie in this region, between the quantum scale and

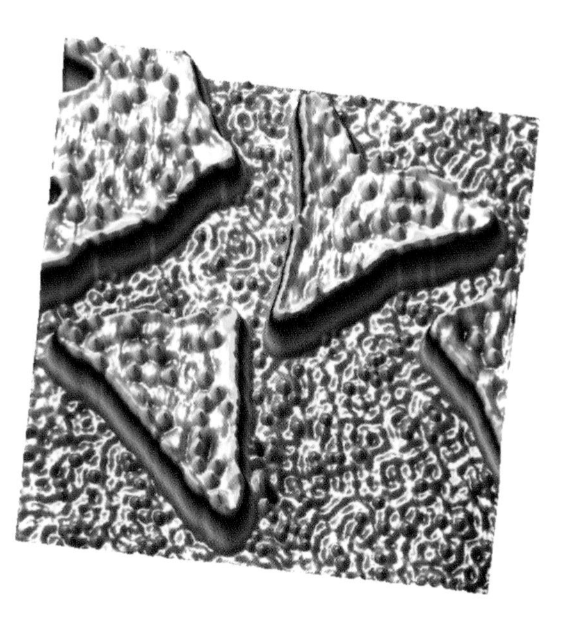

FIGURE 1-6 Spin-polarized scanning tunneling microscope image of magnetic cobalt islands on a Cu(111) surface at T = 4 K. Single atoms of iron have been deposited over the surface (yellow bumps) to study the exchange coupling between single-atom spin centers and magnetic nanostructures. Size of image is 36 × 36 nm. SOURCE: M. Crommie, University of California at Berkeley.

the classical scale, involve AMO science in a number of ways. Nanostructures can be constructed from the bottom up using chemical and optical techniques, or from the top down using techniques such as optical lithography. The structures often have novel optical features, including special absorption properties and negative refractive indices. Nanomaterials with negative indices of refraction could dramatically improve optical microscopes or reduce the feature size in chip fabrication. A new field is growing up to take advantage of these opportunities: nanophotonics.

Size is everything in the nanoworld. The physical, chemical, and biological properties of nanostructured materials can vary substantially at the nanoscale. This dimensional dependence means that physical properties can be controlled by varying the size of the nanoparticles. Until recently, our ability to view or to control the nanoworld was so limited that harnessing it was impossible. However, owing to recent technological developments—many of which are coming from AMO science—efficient and practical nanoscale synthesis and assembly methods will be developed in the coming decade.

Nanofabrication promises to exploit the properties of lasers and optics to improve the production of nanomaterials. Laser ablation is one of the easiest and most widely understood methods for producing nanoparticles from solids. A new twist on this old method is the use of shaped ultrafast pulses to control the size and other characteristics of the nanoparticles. "Atomtronics" is a very new technology that employs trapped ultracold atoms above the surface of a microchip, one aim of which might be to create a single-atom transistor. Nanoscale engineering will allow the creation of new nanostructured media with exceptionally large optical nonlinearities, allowing efficient detection of infrared light. It will also allow the realization of optical nanoparticles, whose size and shape determine their absorption, transmission, and reflection properties. In addition, nanoscale engineering will also allow new optical fibers that can carry the shortest ultrafast pulses without distortion and new lenses that can focus light far more tightly than allowed by the conventional rules of physical optics, and it will allow the construction of new optical displays with unprecedented ruggedness and low cost.

WHAT LIES BEYOND MOORE'S LAW?

Today's computers are doubling in performance every year or two. This will end when the ever-shrinking size of electronic components approaches the level of individual molecules and atoms. While it is still uncertain whether a working large-scale "quantum computer" (as we understand the word computer today) will ever be built, it is clear that quantum mechanics offers a radically different approach to information processing, in which single atoms and photons would be the new hardware. This could lead to computers capable of solving problems that are intractable on any imaginable extension of today's computers but that are important in areas ranging from basic science to national security. Quantum communication might provide security against interception beyond anything possible in today's cyber infrastructure. These applications are based on the strangest and least intuitive concepts of quantum physics, such as Einstein's "spooky action at a distance," which allows "teleportation," or the transfer of information (as opposed to actual physical objects) between remote quantum systems without any physical contact between the quantum hardware during the communication. The possibility of quantum computing is forcing us to explore both theoretical and experimental quantum mechanics at their deepest levels. Should quantum computers be realized, they would be more different from today's high-speed digital computers than those machines are from the ancient abacus. These opportunities are described in Chapter 7 and summarized briefly below.

Quantum mechanics and information theory were two of the scientific cornerstones of the 20th century. One describes physics at very small scales, from molecules and atoms to electrons and photons; the other is a mathematical description of data communication and storage. With the last decades having witnessed the remarkable shrinking of electronic components that carry and process information to near-atomic scales, these two disciplines are naturally beginning to merge. Moore's law of exponentially shrinking computer chip components will soon slow as individual electronic transistors approach the atomic scale, where there is no more room for packing additional components. However, the revolutionary principles of quantum mechanics could offer a way out. Quantum information science may have profound and far-reaching relevance to economic growth, secure communication, and specialized number-crunching. The quantum hardware now found in AMO systems is a key to realizing future quantum devices and will be crucial to the understanding and development of quantum hardware in complex condensed matter systems.

Quantum mechanics contains radical features not found in any other physical theory. The quantum mechanical concept of superposition, where objects can exist in many states simultaneously, is at center stage. When multiple systems are prepared in certain types of "entangled" superpositions, there is a linkage between the systems that does not involve any apparent physical interaction. Einstein called this "spooky action at a distance," and it is the key to the information processing power of quantum information science. The binary digits or bits from conventional information theory now take the form of quantum bits ("qubits"), which can store and process superpositions of numbers in a way that is impossible in any conceivable conventional information processor.

Quantum information theory is a young and rapidly developing field, spanning many areas of science and engineering. Conventional techniques such as logic gate families and error-correction are being adapted to the quantum realm. The landscape of possible quantum applications is still evolving. The best known application is Shor's quantum factoring algorithm, which uses a quantum computer to factor a large number exponentially faster than any known classical algorithm. This has far-reaching implications in the world of cryptography, where most public-key current encryption standards are based on the inability to factor large numbers efficiently. The availability of a quantum factoring machine would render obsolete most of today's encryption standards.

Quantum mechanics offers a remarkable new method for secure data transmission. Quantum cryptography exploits the fundamental tenets of quantum mechanics to allow the secure transmission of information with no physical possibility of undetected eavesdropping. Quantum cryptographic instruments are already available commercially, featuring the use of small numbers of photons traversing a length of optical fiber. There is a rich array of other quantum communication protocols that allow the movement and networking of data in ways that are more efficient than corresponding classical procedures.

AMO physics is concerned with the control and manipulation of atoms, molecules, and photons and is therefore well placed for the development of quantum hardware. Individual atoms confined with electromagnetic fields can be laser-cooled to be nearly motionless and to act as ideal qubit carriers of quantum information (see Figure 1-7). These atoms can be linked by implementing quantum logic gates through direct atom-atom interactions or through individual photons that couple atoms. In this way, large-scale entangled superpositions can be prepared. The use of atomic ion traps, optical lattices, and photons confined between closely spaced mirrors are but a few of the systems that are just starting to show promise for use as future quantum devices.

The grand challenge of quantum information science is the scaling of these AMO systems to the quantum control of even more complex systems. In the realm of condensed matter systems, both superconducting devices that support quantized levels of currents or charges and spin-based devices are now being developed to show rudimentary quantum operations akin to their AMO cousins. While the development of quantum computing and communications hardware currently focuses on AMO science, the future of quantum information science will involve an exciting confluence of scientists and engineers of all stripes.

FIGURE 1-7 At NIST, an ultraviolet laser beam is used to manipulate ions in a high-vacuum apparatus containing an ion trap. These devices have been used to demonstrate the basic operations required for a quantum computer and for teleportation of the quantum state of one atom onto another. SOURCE: National Institute of Standards and Technology, copyright Geoffrey Wheeler.

AMO SCIENCE AND NATIONAL POLICIES: CONCLUSIONS AND RECOMMENDATIONS

The key future opportunities for AMO science contained in this report are based on rapid and astounding developments in the field that are a result of investments made by the federal government's R&D agencies in the work of AMO researchers. In summary, the research field of AMO science and technology is thriving, and the committee offers the following conclusions on the status of the science:

- **Revolutionary new methods to measure space and time have emerged within the last decade from a convergence of technologies in coherent control of ultrafast lasers and ultracold atoms. This new capability creates unprecedented new research opportunities.**
- **Ultracold AMO physics was the most spectacularly successful new AMO research area of the past decade and led to the development of coherent quantum gases. This new field is poised to contribute significantly to the resolution of important fundamental problems in condensed matter science and in plasma physics, bringing with it new interdisciplinary opportunities.**
- **High-intensity and short-wavelength sources such as new x-ray free-electron lasers promise significant advances in AMO science, condensed matter physics and materials research, chemistry, medicine, and defense-related science.**
- **Ultrafast quantum control will unveil the internal motion of atoms within molecules, and of electrons within atoms, to a degree thought impossible only a decade ago. This capability is sparking a revolution in the imaging and coherent control of quantum processes and will be among the most fruitful new areas of AMO science in the next 10 years.**
- **Quantum engineering on the nanoscale of tens to hundreds of atomic diameters has led to new opportunities for atom-by-atom control of quantum structures using the techniques of AMO science. Compelling opportunities in both molecular science and photon science are expected to have far-reaching societal applications.**
- **Quantum information is a rapidly growing research area in AMO science and one that faces special challenges owing to its potential application for data security and encryption. Multiple approaches to quantum computing and communication are likely to be fruitful in the coming decade, and open international exchange of people and information is critical in order to realize the maximum benefit.**

The compelling research challenges embodied in these conclusions are discussed in more detail in the following chapters, which also highlight the broad impact of AMO science on other branches of science and technology and its strong coupling to national priorities in health care, economic development, the environment, national defense, and homeland security.

The linkages to national R&D goals are clear. The White House set forth the country's R&D priorities in the July 8, 2005, memorandum of the science advisor to the President and the director of the Office of Management and Budget. These priorities were reiterated and strengthened in the President's State of the Union Address on January 31, 2006, and in the President's Budget Request for FY2007. AMO scientists contribute to these national priorities in several key areas:

- Advancing fundamental scientific discovery to improve the quality of life.
- Providing critical knowledge and tools to address national security and homeland defense issues and to achieve and maintain energy independence.
- Enabling technological innovations that spur economic competitiveness and job growth.
- Contributing to the development of therapies and diagnostic systems that enhance the health of the nation's people.
- Educating in science, mathematics, and engineering to ensure a scientifically literate population and qualified technical personnel who can meet national needs.
- Enhancing our ability to understand and respond to global environmental issues.
- Participating in international partnerships that foster the advancement of scientific frontiers and accelerate the progress of science across borders.
- Contributing to the mission goals of federal agencies.

An essential part of maintaining the country's leadership in AMO science, and one of the White House's R&D priorities, is to train and to equip the next generation of American scientists. The committee has compiled data on funding, demographics, and program emphasis from the federal agencies to help it assess the current state of AMO science in the United States. In summary, the committee offers 10 more conclusions, this time on government support for AMO science:

- **Given the budget and programmatic constraints, the federal agencies questioned in this study have generally managed the research profile of their programs well in response to the opportunities in AMO science. In doing so, the agencies have developed a combination of modalities (large groups; centers and facilities; and expanded single-investigator programs). Much of the funding increase that has taken place at DOE, NIST, and NSF has been to benefit activities at research centers. The overall balance of the modalities for support of the field has led to outstanding scientific payoffs.**
- **The breadth of AMO science and of the agencies that support it is very**

important to future progress in the field and has been a key factor in its success so far.

- Since all of the agencies report that they receive many more proposals of excellent quality than they are able to fund, it is clear that AMO science remains rich with promise for outstanding future progress. AMO science will continue to make exceptional advances in science and in technology for many years to come.
- In view of its tremendous importance to the national well-being broadly defined—that is, to our nation's economic strength, health care, defense, education, and domestic security—an enhanced investment program in research and education in physical science is critical, and such a program will improve the country's ability to capture the benefits of AMO science.
- Historically, support for basic research has been a vital component of the nation's defense strategy, making the recent decline in funding for basic research at the defense-related agencies particularly troubling.
- The extremely rapid increase in technical capabilities and the associated increase in the cost of scientific instrumentation have led to very significant added pressures (over and above the usual Consumer Price Index inflationary pressures) on research group budgets. In addition, not only has the cost of instrumentation increased, but also the complexity and challenge of the science make investigation much more expensive. This "science inflator" effect means that while it is now possible to imagine research that was unimaginable in the past, finding the resources to pursue that research is becoming increasingly difficult.
- In any scientific field where progress is extremely rapid, it is important not to lose sight of the essential role played by theoretical research. Programs at the federal agencies that support AMO theory have been and remain of critical importance. NSF plays a critical and leading role in this area, but its support of AMO theoretical physics is insufficient.
- AMO science is an enabling component of astrophysics and plasma physics but is not adequately supported by the funding agencies charged with responsibility for those areas.
- The number of American students choosing physical science as a career is dangerously low. Without remediation, this problem is likely to create an unacceptable "expertise gap" between the United States and other countries.
- Scientists and students in the United States benefit greatly from close contact with the scientists and students of other nations. Vital interactions include the training of foreign graduate students, international

collaborations, exchange visits, and meetings and conferences. These interactions promote excellent science, improve international understanding, and support the economic, educational, and national security needs of the United States.

Finally, the committee offers the following six recommendations, which as a whole form a strategy to fully realize the potential at the frontiers of AMO science.

Recommendation. In view of the critical importance of the physical sciences to national economic strength, health care, defense, and domestic security, the federal government should embark on a substantially increased investment program to improve education in the physical sciences and mathematics at all levels and to strengthen significantly the research effort.

Recommendation. AMO science will continue to make exceptional contributions to many areas of science and technology. The federal government should therefore support programs in AMO science across disciplinary boundaries and through a multiplicity of agencies.

Recommendation. Basic research is a vital component of the nation's defense strategy. The Department of Defense, therefore, should reverse recent declines in support for 6.1 research at its agencies.

Recommendation. The extremely rapid increase in the technical capability of scientific instrumentation and its cost has significantly increased pressures (over and above the usual Consumer Price Index inflationary pressures) on research budgets. The federal government should recognize this fact and plan budgets accordingly.

Recommendation. Given the critical role of theoretical research in AMO science, the funding agencies should reexamine their portfolios in this area to ensure that the effort is at proper strength in workforce and funding levels.

Recommendation. The federal government should implement incentives to encourage more U.S. students, especially women and minorities, to study the physical sciences and take up careers in the field. It should continue to attract foreign students to study physical sciences and strongly encourage them to pursue their scientific careers in the United States.

2

AMO Science and the Basic Laws of Nature

Atomic, molecular, and optical (AMO) experiments have reached such high levels of sophistication, precision, and accuracy that they are uniquely positioned to carry out the most demanding tests ever conducted of some of the most fundamental laws of nature. For example, AMO scientists aim at measuring subtle new effects due to a possible permanent electric dipole moment of an electron or an atom, which, if such exists, would require a dramatic extension of our theory of elementary particles. But this measurement proceeds via ultraprecise measurements in atomic physics rather than by using high-energy particle accelerators. Atomic physics experiments can also search for violations of the so-called charge, parity, and time reversal (CPT) symmetry, which states that matter and antimatter particles should have exactly the same mass and the same magnetism. Both of these symmetries are fundamental pillars of modern physics, so testing them is of the greatest significance. In yet another development, theories unifying gravity with the other fundamental interactions of nature suggest the possibility of spatial and temporal variations of physical "constants" such as the fine structure constant α of electromagnetism. In addition to being used to measure α to unprecedented precision as a test of quantum electrodynamics, techniques from AMO science can be used to place limits on the variation of α over cosmological timescales. And laser-based gravitational-wave detectors will characterize gravitational waves and unravel information about their violent origins and about the nature of gravity. These and other applications of AMO science for exploring the basic laws of nature are the topics of this chapter, along with some of the practical applications that result from this quest.

SPIN SCIENCE

One of the most interesting phenomena in quantum mechanics is that nature's subatomic particles possess an intrinsic property called "spin." First observed by Stern and Gerlach in 1922 (see Box 2-1), this property has no exact analogy in the classical world. For while planets may spin on their axes, the property is not an in-

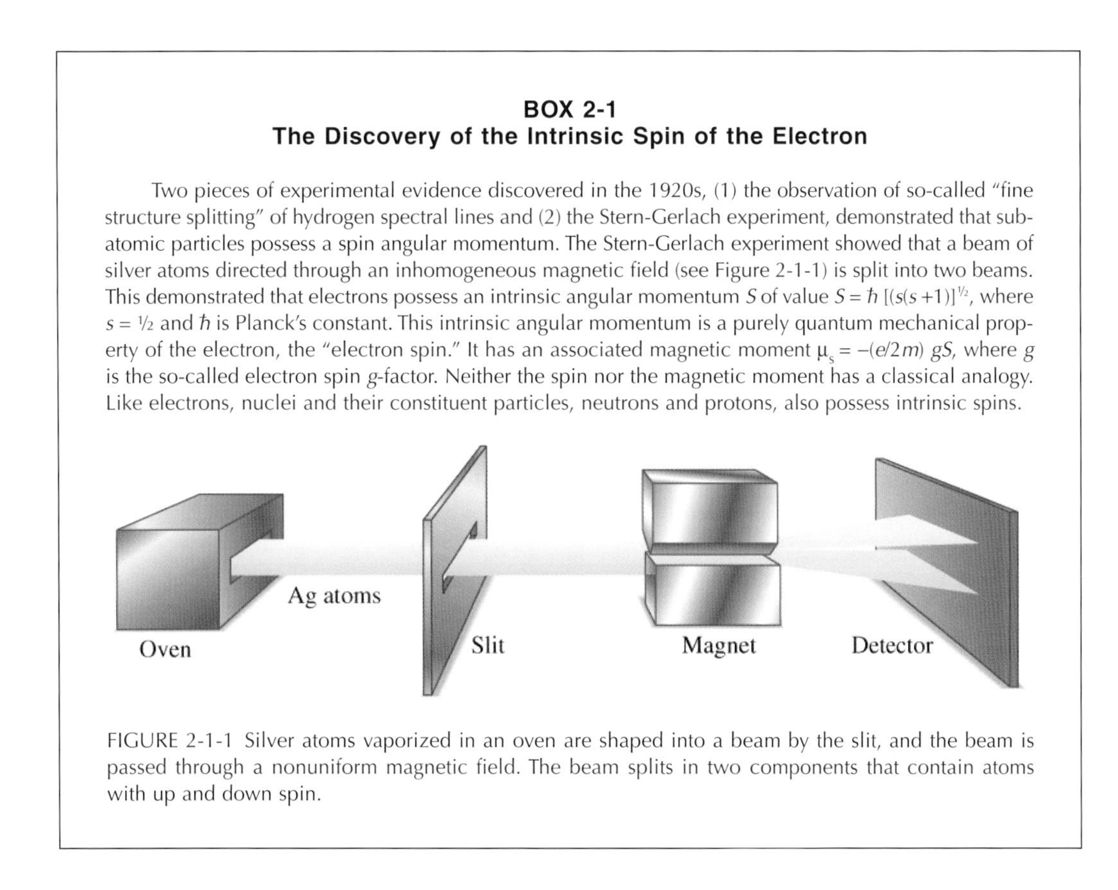

BOX 2-1
The Discovery of the Intrinsic Spin of the Electron

Two pieces of experimental evidence discovered in the 1920s, (1) the observation of so-called "fine structure splitting" of hydrogen spectral lines and (2) the Stern-Gerlach experiment, demonstrated that subatomic particles possess a spin angular momentum. The Stern-Gerlach experiment showed that a beam of silver atoms directed through an inhomogeneous magnetic field (see Figure 2-1-1) is split into two beams. This demonstrated that electrons possess an intrinsic angular momentum S of value $S = \hbar\,[(s(s+1)]^{1/2}$, where $s = 1/2$ and $\hbar$ is Planck's constant. This intrinsic angular momentum is a purely quantum mechanical property of the electron, the "electron spin." It has an associated magnetic moment $\mu_s = -(e/2m)\, gS$, where g is the so-called electron spin g-factor. Neither the spin nor the magnetic moment has a classical analogy. Like electrons, nuclei and their constituent particles, neutrons and protons, also possess intrinsic spins.

FIGURE 2-1-1 Silver atoms vaporized in an oven are shaped into a beam by the slit, and the beam is passed through a nonuniform magnetic field. The beam splits in two components that contain atoms with up and down spin.

trinsic one—each planet in general does it in a different way. In the quantum world however, each elementary particle of a given type has exactly the same intrinsic spin angular momentum. This spin of the particle is also directly related to the intrinsic magnetism of the particle. In turn, that magnetism is one of the main ways in which the particle interacts with other particles and with its environment.

The magnetic forces on the spins of electrons and nuclei in atoms can be measured with remarkable precision because in many atoms the spins couple only weakly to other atoms in the environment. In fact, spin measurements are among the most precise in all of science. But sometimes it is the interactions with the environment itself that are interesting, as in magnetic resonance imaging (MRI), and atomic spin science is contributing to advances in that area. Other times, it is a possible new coupling of spins that is of interest. For example, it is predicted by the theory of supersymmetry that electrons and nuclei in atoms should possess a tiny offset between their center of electric charge and their center of mass, and that this offset should lie along the axis of spin. This shift is called a permanent electric dipole moment (EDM). If it exists, it would be caused by a force that violates a cherished law in physics: time reversal (T) invariance (see Figure 2-1).

T-violating forces in the conventional Standard Model of elementary particle interactions generate an EDM far too small to be observed by any presently envisioned experiment, so the discovery of an EDM would reveal new forces and particles that lie outside the Standard Model. Such new forces and particles are absolutely needed, because the Standard Model has some glaring gaps. For example, it does not predict the overwhelming preponderance of matter over antimatter in the universe. But there is very little experimental evidence that would help establish any one of the alternative theories. According to some of the favorite contenders, such as supersymmetry, new forces should exist and should produce atomic EDMs at measurable levels. Figure 2-2 shows the ranges of some of the current predictions for EDMs arising from various dynamical models that extend the Standard Model of particle physics.

Since the energy of an atom with an EDM will depend upon the direction of the spin axis relative to an applied electric field, EDM experiments look for a tiny shift in the nuclear or electron spin resonance of an atom or molecule placed in a large electric field. Such spin resonance frequency measurements may provide our best opportunity to detect these tiniest effects of new fundamental forces that cannot yet be seen in experiments at high-energy accelerators. Atomic physics is taking up this challenge by creating ultrasharp spin resonances associated with long spin stability (hundreds to thousands of seconds in the case of nuclear spins in atomic gases or vapors). Experiments can now detect frequency shifts smaller than a nanohertz—less than one complete spin precession in 30 years. At such sensitivity,

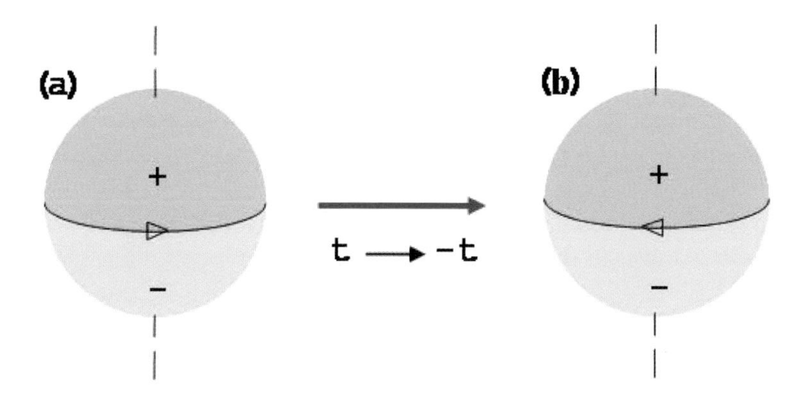

FIGURE 2-1 An electric dipole moment (EDM) of an atom (*left*) is a permanent separation between the centers of positive and negative charge along the axis of spin. Under time reversal (*right*), the spin direction is reversed but the charge separation is not. An observed EDM would have to be caused by forces that violate time reversal symmetry.

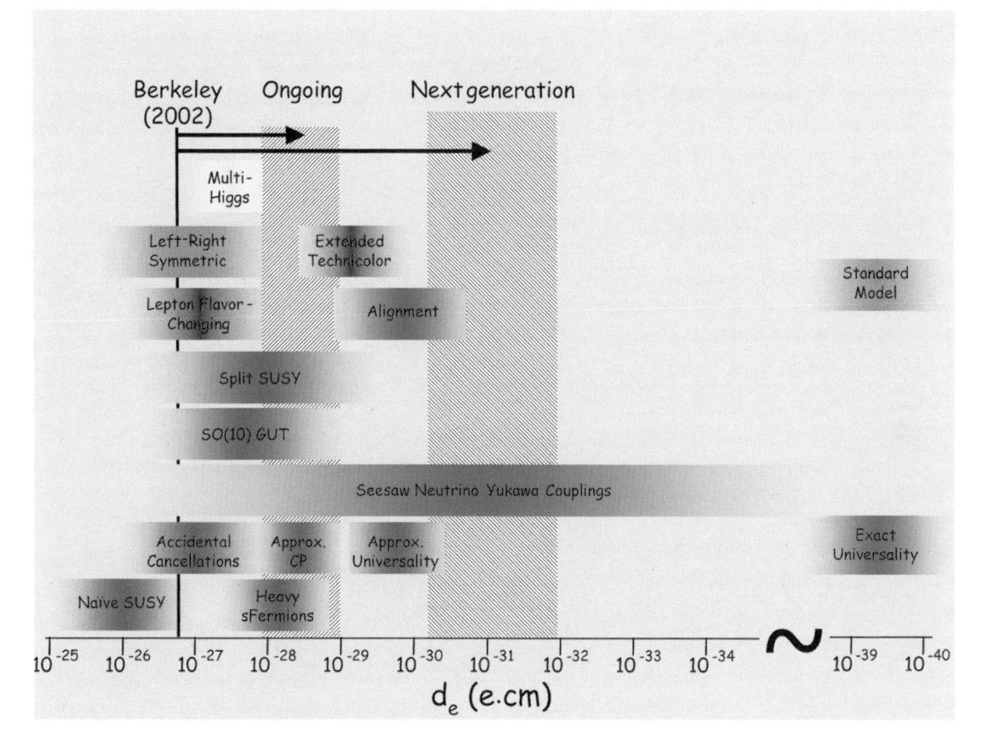

FIGURE 2-2 Scientific impact of current and next-generation electron EDM measurements. It is anticipated that next-generation measurement will reach the 10^{-31} e-cm level, equivalent to a dipole consisting of a positive and negative electronic charge separated by only 10^{-31} cm, which will test large classes of supersymmetric (SUSY) theories. Ongoing measurements of an atomic EDM due to nuclear spin and the EDM of the neutron have similar sensitivity to different SUSY parameters. Thus, atomic EDM experiments will be probing SUSY and other theories on a broad front. SOURCE: D. DeMille, Yale University.

subtle new effects due to EDMs could be seen if they are present (see Box 2-2). At the same time, such sensitivity to spins and to the effect of magnetic fields on spins is opening up new applications in medical research and diagnosis.

Magnetometry and Medical Imaging

Atomic spin magnetometers, developed in part for fundamental EDM and Lorentz experiments, are now surpassing low-temperature superconducting quantum interference devices (SQUIDs), formerly the industry standard for precise magnetometry. Atoms have higher sensitivity, higher spatial resolution, and no

BOX 2-2
Using Atoms to Probe the Particle Physics Frontier

At first sight, it is remarkable that the study of atoms can tell us about forces that are normally explored at high-energy accelerators or large underground labs. Not only are atomic experiments carried out at ordinary energies, usually on a tabletop scale, but inside atoms the electrons are bound by well-understood electric and magnetic forces and move at very small energies compared with those reached at particle accelerators. Indeed, in much of atomic physics we can consider bound atomic electrons as point charges interacting with the electric and magnetic fields of the other charges in the atom. Close to an electron, however, there is a lot more happening. The electron envelops itself in a tiny cloud of emitted and reabsorbed virtual particles (Figure 2-2-1)—not only photons, but even the heavy particles studied at high-energy accelerators (similarly, the elementary constituents of the nucleus, the quarks, are surrounded by even more complicated clouds). The existence of these tiny clouds shows up in highly sensitive measurements of energy levels and other atomic properties, such as the exquisite measurements of the magnetism of the electron discussed in the next subsection, "Magnetometry and Medical Imaging." Furthermore, the clouds will contain all virtual particles, even those too heavy to be produced by accelerators, so new physics beyond the current Standard Model might well show up first in atomic experiments. For example, it is widely believed that there are so-called supersymmetric particles that will eventually be found in high-energy experiments at new facilities such as the Large Hadron Collider (LHC). But for now the atomic EDM experiments are well positioned to discover supersymmetric particles if they do indeed exist.

In fact, other atomic experiments have already shown how the small virtual clouds around each electron and quark endow atoms with remarkable properties that are of great interest to particle physics. One particle in such clouds, the 90 GeV Z^0 boson, distinguishes left from right (that is, it violates parity) and causes an isolated atom to absorb right-handed, circularly polarized photons at a slightly different rate than left-handed photons. The effect is very small, but the measurements in the cesium atom are so precise and agree so well with the Standard Model as to show that a proposed heavier version of the Z^0 must be at least eight times heavier than the known Z^0, if it exists at all. Thus atomic physics, through parity violation measurements, has been able to give particle theorists information long before particle accelerators will have attained sufficient energy (700 GeV) to test for the existence of such a particle.

need for cryogenic apparatus. Developments are under way that will allow measurements of the weak magnetic fields produced by the brain and the heart, which can provide valuable diagnostic and research information. These techniques are expected to help in diagnosis of epilepsy, cardiac arrhythmias, and other diseases. They also can be used in functional studies of the brain.

Advances in measuring nuclear spin resonances in the case of noble gases not only are helping in the searches for an EDM and Lorentz violation but are also opening up a new field in medical imaging. Recently, high-quality magnetic resonance images based on the gases ^{129}Xe and ^{3}He have begun to appear. For example,

FIGURE 2-2-1 Each atomic electron (and quark) is surrounded by tiny clouds of virtual particles. The greater the mass of a virtual particle, the smaller its cloud. The Z^0 boson, for example, produces a cloud about one ten-millionth the size of an atom. Precise atomic experiments have measured the virtual particles' effect to better than 1 percent.

as illustrated in Figure 2-3, the image of a human lung momentarily filled with ^{3}He gas provides a high-resolution picture of the lung. Ordinary MRI techniques are not capable of producing such pictures. The clinical result is that lung disease can be diagnosed much earlier with these new instruments.

The secret to obtaining such images is to increase the fraction of nuclear spins that are aligned with each other, so that their magnetism adds up to a larger value. In ordinary MRI, this fraction is only a few parts per million. That is due to the trade-off between the aligning force exerted on the magnetic moments of the atoms by a strong magnetic field and the randomizing force the molecules exert on

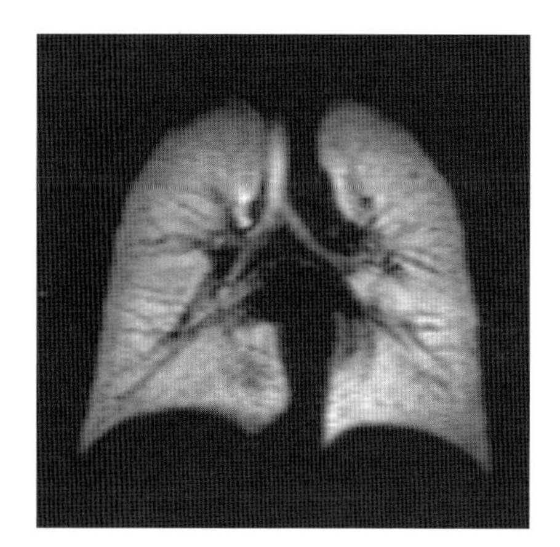

FIGURE 2-3 A coronal image of the lungs using magnetic resonance, when the lungs are momentarily filled with hyperpolarized atomic ^{3}He gas. SOURCE: Center for In-vivo Hyperpolarized Gas MRI, Radiology Department, University of Virginia.

each other. But polarized laser light can pump angular momentum into the atoms very efficiently through a process of repeated optical absorption and fluorescence, called "optical pumping." In this way, the alignment can be increased to several tens of percent. For such optically pumped, hyperpolarized atomic gases, the signal per nucleus is about 100,000 times larger than the signal per nucleus in normal MRI. This more than makes up for the smaller density of the nuclei in these gases compared with the density in human tissue.

Spin and Basic Forces

One effect of the cloud of virtual particles around the electron illustrated in Figure 2-2-1 is to modify the magnetism associated with the electron spin by about 0.1 percent. Using a single electron held in a magnetic trap, this tiny modification has been measured to astonishing accuracy: 10 decimal places. Equally astonishing, the theory of these radiative corrections has been used to calculate the effect to the same precision. The calculated result is expressed in terms of the fine structure constant α, which gives the strength of the electric coupling. It remains now for other atomic experiments, discussed in the subsection "Fine Structure Constant," to measure α to equal precision; then it will be known whether or not the theory of quantum electrodynamics agrees with experiment to such unprecedented accuracy. In a more speculative vein, atomic experiments are so sensitive to the very fabric of space and time that they can measure its underlying symmetries as embodied in the principle of Lorentz invariance. This is the principle of relativity put forward by

Einstein in 1905. Atomic experiments can also search for violations of the so-called CPT symmetry, which states that matter and antimatter particles should have exactly the same mass and the same magnetism. Both of these symmetries are fundamental pillars of modern physics. A recently developed theoretical framework now exists for linking a possible Lorentz violation with CPT violation. Precise Lorentz tests include spin resonance experiments in which the spin precession frequency would depend on the orientation of the precession axis relative to a fixed direction in space, the vector $\mathbf{b}_i$ in Figure 2-4. One of the most exact CPT tests is expected to result from a remarkable recent feat, the creation of antihydrogen in the laboratory. By comparing the spectral energies of antihydrogen atoms and ordinary hydrogen atoms, one can search for effects due to CPT violation.

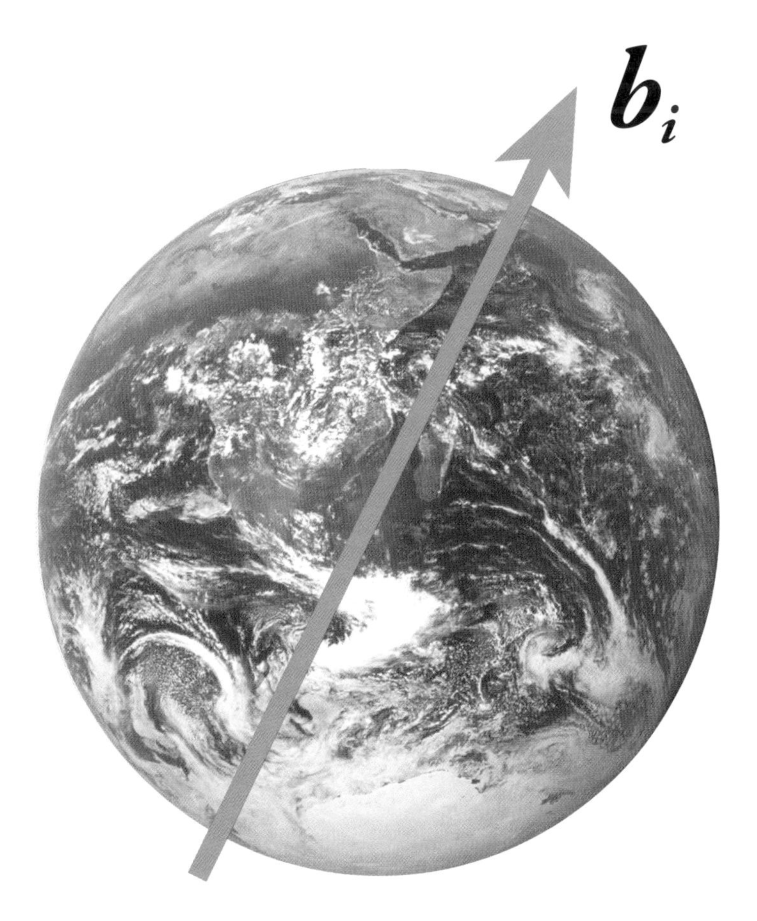

FIGURE 2-4 If there is a preferred direction $\mathbf{b}_i$ in space in the vicinity of Earth, sensitive spin experiments will show a diurnal effect on Earth as it rotates. SOURCE: Image of Earth, National Aeronautics and Space Administration.

The great challenge in most EDM experiments and Lorentz symmetry tests is to detect the tiniest possible shifts in spin resonance frequency. The EDM sensitivity is mind-boggling. In the case of the mercury atom, if the atom were blown up to the size of Earth, the separation between centers of plus and minus charge would be less than one-hundredth of an angstrom!

ENERGY LEVELS, TIME, AND ATOMIC CLOCKS

From the discovery of the laws of planetary motion in the Renaissance to the development of quantum theory in the 20th century, precision timekeeping has historically accompanied advances in science. In particular, advances in AMO science have led to ever more accurate atomic clocks. Likewise, improvements in clocks have enabled revolutions in technology, starting with the determination of longitude in the 17th century and leading to the development of the modern global positioning satellite (see Figure 2-5 for one such system).

The measurement of frequency (or, equivalently, time duration) can be done with greater precision than any other measurement in physics. This precision stems from the ability to compare signals very accurately to natural periodic phenomena, such as planetary motion, the swinging of a pendulum, or the vibrations in a crystal of quartz. The pinnacle of frequency standards is the atomic clock, which relies for its precision on the generation of a highly accurate oscillating signal based on the vibrations in an isolated atom. This oscillation comes from the difference in energy ΔE between two states in the atom and the relationship of the energy difference to frequency, ν, through Planck's law, $h\nu = \Delta E$, where h is Planck's constant of quantum mechanics. The frequency of an atomic clock thus depends only on the properties of an isolated atom, regardless of whether that atom is located in Washington, D.C., or Beijing. The extraordinary advances in atomic clocks over the past 15 years have resulted mainly from isolating and cooling ions or neutral atoms, thereby eliminating Doppler shifts and other perturbations (see Box 2-3). The state-of-the-art accuracy of atomic clocks, which is approaching one second in 60 million years, involves the use of a small number of neutral atoms or a single ion in electric traps. Such accuracies are not merely academic—as indicated above, they are necessary for increasing the precision of navigation, from applications in space exploration to advanced architectures in GPS.

To attain such accuracy, the atoms in atomic clocks must be isolated from their environment, including from other nearby atoms. For this reason, accurate atomic clocks usually deal with very small numbers of atoms, in some cases just a single atom. However, extracting any useful information from such a small number of atoms requires collecting and averaging the weak signals they emit over very long periods of time. Even the most efficient detector will see at most one photon every

FIGURE 2-5 Schematic drawing of the Global Positioning System (GPS) constellation. The GPS uses 24 satellites (21 plus spares) at an altitude of about 11,000 miles, moving at about 7,000 mph. A GPS receiver on Earth synchronizes itself with the satellite code and measures the elapsed time since transmission by comparing the difference between the satellite code and the receiver code. The greater the difference, the greater the time since transmission. Knowing the time and the speed of light, the distance can be calculated. The time comes from four atomic clocks on each satellite. The clocks are accurate to within 0.003 seconds per 1,000 years. Using information from four or more satellites, the GPS receiver calculates latitude, longitude, and altitude. SOURCE: National Aeronautics and Space Administration.

time an atom decays, since only one photon is given off in the decay process. To find the decay rate, we have to count for a long time. This rate measurement is precisely the same kind of problem as measuring the odds for heads or tails in a coin toss. You have to play a long time to get beyond the small statistics of winning or losing streaks, to find the true average rate for tails.

Specifically, for the coin toss or any other random classical counting process, N trials are needed to produce a precision of $1/N^{1/2}$. This is known as the shot-noise limit. But shot noise is one of those seemingly immutable facts of nature that need to be rethought in the quantum realm of atoms. Indeed, one of the applications of

BOX 2-3
History of Laser-Cooled Atoms and Ions

Over 30 years ago, in 1975, two pairs of West Coast scientists proposed the remarkable idea that a gas of atoms could be cooled by illuminating it with laser light. Only a few other scientists took much immediate note of this new idea, but by the late 1970s it had been successfully applied to cooling electrically charged and trapped ions to very cold temperatures, a thousandth of a degree above absolute zero. By the early 1980s other groups were trying to laser cool neutral atoms. Crucial support for this early work came from the Office of Naval Research (ONR), motivated by the hope that laser cooling would lead to better clocks, an essential element of modern precision navigation. Since that time, laser cooling has led to the realization of Bose-Einstein condensation in dilute gases and the achievement of the coldest temperatures known in the universe—less than a billionth of a degree above absolute zero. Eight atomic physicists have received Nobel prizes for work related to laser cooling and trapping, and hundreds of research groups are devoted to its study. Today, cold atoms represent a large fraction of all the research in AMO science. The frequency standards of the world's major standards laboratories are laser-cooled atomic clocks (see Figure 2-3-1). International time ticks at a rate controlled by laser-cooled atoms. And the U.S. Naval Observatory, reaping the rewards of the ONR's investment, is installing an ensemble of laser-cooled clocks to keep time for the nation's military.

Cold atom physics continues to bear fruit: Major research programs on precision inertial navigation systems are under way in the United States and abroad; quantum degenerate Bose and Fermi atomic gases are modeling condensed matter systems of interest for both basic science and practical devices (Chapter 3); and ions and atoms are being developed as the qubits for the new science of quantum information (see Chapter 7).

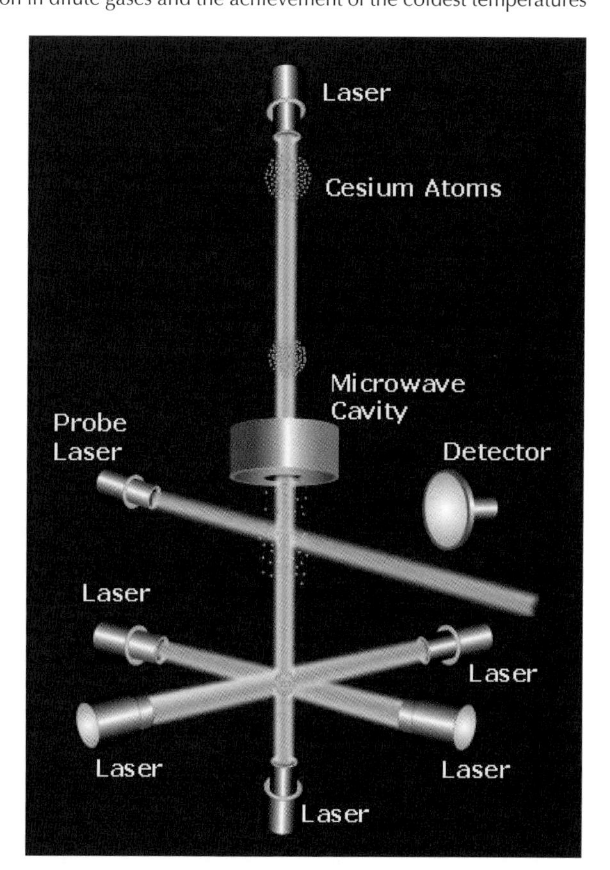

FIGURE 2-3-1 Layout of the NIST-Boulder atomic fountain laser-cooled clock NIST F-1. This clock is now the most accurate clock in the world, ticking off time with an uncertainty of a few parts in 10^{16}, about 1 second in 60 million years. SOURCE: J. Aarons, National Institute of Standards and Technology.

the special feature of quantum mechanics known as "entanglement" or "quantum correlation" is to increase the amount of information that can be extracted from a single measurement of many atoms, effectively beating the shot-noise limit. By entangling N atoms in an atomic clock, it is possible for the inaccuracy in the averaged measurement to be improved to $1/N$ instead of $1/N^{1/2}$. This level of precision, the best allowed by quantum mechanics, is called the Heisenberg limit. For even a small number of atoms (say, $N = 1{,}000$), this improvement is substantial. The recipe for operating an atomic clock in the Heisenberg limit is to create a collective quantum state of all of the atoms, sometimes called a "superatom," that effectively speeds up the atomic clock by a factor of N.

New Clock Technologies and GPS

Precision oscillators and clocks have far-reaching, economically significant applications, including network synchronization, high-performance analog-to-digital conversion, and global positioning. It has been recognized for decades that specialized laser light sources have attributes that enable breakthrough performance for timing applications. These attributes include exceptional frequency stability and accuracy. However, these attributes could not be exploited for applications since it was not known how to transfer this stability and accuracy from the optical frequency range (10^{15} Hz) to frequencies accessible with traditional electronics ($<10^{10}$ Hz). This long-standing problem, known as optical frequency counting, found a beautiful and technologically compelling solution by exploiting features associated with the output of femtosecond pulsed laser sources. The physicists who pioneered this breakthrough were awarded the 2005 Nobel prize in physics. As a result of efforts in many laboratories over the past 5 years, it is now possible to translate frequency stability and accuracy from the optical regime to the microwave regime and also to measure the ratio of optical frequencies with unprecedented precision. Very recent work indicates frequency counting methods may be extensible into the x-ray regime. This has direct, near-term technological impact. For example, the performance of high-end analog-to-digital converters—the key component in advanced radar systems—is currently limited by the frequency stability of current-generation clocks. Next-generation optical oscillators enable clock jitter at the 1 femtosecond level, which is expected to improve the system performance in terms of both bandwidth and signal-to-noise ratio.

The atomic clock plays a central role in GPS architecture. Future advances in GPS performance will be enabled by high-stability clocks in the receiver and also in the satellite. Very recently, microfabrication techniques have been employed to create miniature (so-called "chip scale") clocks with breakthrough performance characteristics for the receiver clock (Figure 2-6). Similar miniaturization efforts are

FIGURE 2-6 *Left:* Miniature atomic clock occupying less than 10 mm^3. *Right:* The U.S. primary frequency standard, based on laser-cooled cesium atoms. SOURCE: National Institute of Standards and Technology.

being applied to produce compact magnetometers and gyroscopes. It is expected that these miniature atomic clocks will be widely deployed. Accurate atomic clocks also play important roles in submarine navigation systems.

Are the Constants of Nature Changing?

Our theoretical understanding of the universe and the known forces of nature incorporates a number of fundamental constants. Most famously, the speed of light, *c*, in a vacuum is constant to all observers. Other no less important physical constants are the fundamental quantum of angular momentum, known as Planck's constant, *h*, and the charge *e* of the electron. From these three constants for charge, angular momentum, and speed we can form a dimensionless ratio (specifically, $e^2/\hbar c$),[1] known as the fine structure constant α. Although gravity was long thought to be truly constant over the lifetime of the universe, recent theories unifying it with the other physical interactions suggest the possibility of spatial and temporal variations of α and other physical "constants."

Atomic physics comes into this picture because different atomic transitions depend differently on α so that comparing the rates of different atomic clocks over long periods of time allows one to put bounds on the local change of α with

[1]This formulation of the fine constant uses the term $\hbar$, which is called h-cross or h-bar. $\hbar$ is Planck's constant *h* divided by twice the value of π.

time. Already there is a strong laboratory limit on the time variation of α: less than about one part in 10^{15}. Recently, however, the accuracy achieved by atomic clocks has improved enough to test faint hints from cosmological and early Earth studies that α may have changed very slowly over the lifetime of the universe. Spectroscopic data provide some limited evidence that the value of α may indeed have evolved as the universe expanded. If this is confirmed, our understanding of quantum electrodynamics and its relation to the structure of spacetime would need to undergo major revision.

MEASURING DISTANCE AND MOTION USING INTERFEROMETERS

Interference is the term used in optical science for what happens when two (or more) optical beams that have traveled along different paths from the same source are brought back together. When the wave crests of two beams match up with each other, a bright spot is produced; when the wave crests of one beam match up with the wave troughs of the other, they cancel each other out and produce a dark spot. Because the wavelength of light is so short, about a micron or less, an interference meter (interferometer) is very sensitive to tiny relative changes in path length (or velocity) because of the Doppler effect. As the word "optical" in its name implies, AMO science continues to be at the forefront of the development of optical interferometers. The most celebrated practical application has been the ring laser gyroscope for navigation, which is the present industry standard. Exciting examples of the use of optical interferometry for fundamental physics and astronomy are at the ground-based Laser Interferometer Gravity-Wave Observatory (LIGO) and the proposed space-based Laser Interferometer Space Antenna (LISA), which are designed to detect directly gravitational waves coming from distant points in the universe (see subsection after next).

At the same time, AMO research is leading to new and even more sensitive interferometers based on the much shorter wavelengths associated with moving particles. This very active area is called de Broglie (or matter-wave) interferometry. New instruments based on this technique are expected to revolutionize the fields of inertial navigation and gravitational anomaly detection. There is also a remarkable application of matter-wave interferometry to measuring the fine structure constant α, which could help test the fundamental quantum theory of electricity and magnetism to unprecedented accuracy.

Optical Sensors for Navigation

Workhorse inertial sensors used widely in commercial and military navigation rely on ring-shaped lasers called ring laser gyroscopes (RLGs), which detect rotation by comparing two light beams that travel around the same ring in opposite directions. The difference in light travel time in the two directions is proportional to the product of the rotation rate of the interferometer platform and the area enclosed by the interferometer.

A similar principle underlies the operation of fiber-optic gyroscopes (FOGs), in which the ring is replaced by loops of fiber. Advances in optical fiber and integrated optics technologies have resulted in significant improvements in the performance and cost of FOGs, to the extent that next-generation navigation systems will probably employ FOG technology rather than RLG. Developments in photonic band gap fiber technology may lead to another performance leap for these strategically vital systems. Photonic band gap fibers contain microstructured central cores that permit the light to travel mostly in air rather than glass.

Precision inertial navigation requires a detailed knowledge of Earth's gravitational field (the acceleration due to gravity needs to be subtracted from the accelerometer outputs). Historically, this has resulted in close links between the navigation and geodesy communities. Advances in inertial technology have direct consequences for geodesy. For example, both GPS sensors and a large-area RLG detected minute changes in Earth's rotation rate attributable to the tsunami in the Indian Ocean at the end of 2004. Laser-based gravimeters are being used worldwide to characterize Earth's gravitational field. Recently they have been used for oil well logging and resource management.

Direct Detection of Gravitational Waves

Large-scale optical interferometric sensors form the cornerstone measurement technology for LIGO, the nation's first great gravitational-wave observatory (see Figure 2-7), and for LISA, a future orbiting gravity-wave observatory. Gravitational waves are predicted to be tiny ripples in the otherwise smooth fabric of spacetime produced by violent events in the distant universe—for example, by the merging of two neutron stars or two black holes, or in the cores of supernova explosions. They have never been observed directly, but their influence on the orbital motion of the corotating binary pulsar PSR1913+16 has been confirmed by direct measurement. This observation was celebrated by the award of the Nobel prize in 1993.

Gravitational waves are emitted by accelerating masses much as electromagnetic waves are produced by accelerating charges. They travel to Earth at the speed of light, bringing with them information about their violent origins and about the nature of gravity. To detect gravity waves, which are very weak, requires a huge, highly precise optical interferometer. The LIGO design features two main interferometers, one in Hanford, Washington, and one in Livingston Parish, Louisiana. The interferometers are in the shape of an L with 4-km arms, each of which contains

FIGURE 2-7 Aerial view of one arm of the LIGO gravitational-wave detector in Livingston Parish, Louisiana. Two other large interferometers make up the second LIGO site at Hanford, Washington. SOURCE: LIGO Laboratory.

ultrastable laser beams that bounce back and forth millions of times between two freely hanging test masses fitted with mirrors, one at each free end of an arm. These assemblies are housed within high-vacuum tubes to eliminate light scattering from air. A third freely hanging mirrored test mass hangs at the vertex of the L. When a gravitational wave passes by, the distance between the test masses will change by a different amount in each arm (only about a hundred-millionth of the diameter of a hydrogen atom over the 4-km length of the arm). Such a tiny change can be detected by the change in relative phase between the laser beams in the two arms. A key challenge is to isolate the test masses from other disturbances such as seismic vibrations and air molecules. An upgrade to LIGO is expected to improve the instrument's sensitivity by an order of magnitude.

Advanced measurement and instrumentation techniques developed for LIGO experiments—such as "squeezed light" sources/detectors and thermal noise mitigation strategies—may have practical impact in next-generation commercial and strategic sensors.

Matter Wave Interferometry (de Broglie Wave Interference)

A completely different approach to improved inertial navigation might use matter waves instead of light waves. While the existing generation of navigation sensors perform at superb levels, advances in navigation technology are considered essential to support plans for broadly deployed autonomous defense and communications systems. Matter waves, known as de Broglie waves after the physicist who first proposed the wave-matter connection in quantum theory, might greatly improve inertial system accuracy, performance, and price. These sensors operate on wave-interference principles analogous to those underpinning the operation of the optical sensors described above. Were de Broglie wave sensors to reach the threshold of commercial viability, this could be considered one of the first great applications for ultracold atom physics, discussed in detail in Chapter 3.

Future gravimetry based on de Broglie wave interference will make possible long-distance airborne characterization of gravitational anomalies at unprecedented levels, for accurate detection of underground structures and tunnels as well as buried minerals and other natural resources. Widespread adoption of this technology requires significant progress in the development of supporting technologies, including robust, compact, and cost-effective laser and vacuum systems. Existing prototype systems require extensive vibration isolation in order to maintain the required benign environment for sensor operation. Development of interferometer configurations that mitigate sensitivity to platform motion while sustaining or improving sensor performance remains a significant scientific challenge.

The current generation of atomic sensors is based on laser-cooled ensembles of cold atoms in free space and pulses of laser light. The development of Bose-Einstein condensed/degenerate Fermi gas atomic sources and of de Broglie waveguide methods raises the interesting question whether these methods can be exploited to improve sensor performance. The central scientific challenge associated with this question is to understand how the inertial/gravimetric-phase information associated with the propagation of de Broglie waves is affected by interactions arising from, for example, atom-atom collisions, waveguide imperfections, atom-surface coupling, spurious magnetic and optical fields, and vibrations. Surprisingly, fermionic atoms now appear to be a viable alternative to bosons for such applications (see Chapter 3 for a discussion on the differences between cold bosons and fermions). An advantage of quantum-degenerate fermions for atom interferometry is that they do not suffer the kind of collisions that bosons are subject to, which can rapidly lead to decoherence and the destruction or random shift of an interference pattern. This is a direct consequence of the Pauli exclusion principle, which forbids two fermions to occupy the same quantum state. This principle also renders fermionic interferometry somewhat analogous to white-light interferometry in conventional optics.

Fine Structure Constant

Matter-wave interference experiments can be used to make precision measurements of the value of the fine structure constant, α. The measurement of this constant is not just a curiosity but is an important component of critical tests of the quantum theory of electromagnetism, quantum electrodynamics (QED). To obtain α, a number of experiments that evolved from developments in different communities—laser cooling, de Broglie wave interferometry, and precise optical and mass spectroscopy—must be combined. The current goal is to measure α to a part in a billion accuracy. This value can then be used to calculate the spin magnetism of the electron, which—as discussed in the subsection "Optical Sensors for Navigation"—has already been measured to such accuracy. The result will be a test of our understanding of QED to a part per billion.

AMO PHYSICS IN THE STUDY OF THE DISTANT UNIVERSE

Early in the last century astronomers discovered that the light from distant stars and galaxies exhibits characteristic spectral features that can be accurately predicted using the atomic physics theory gleaned from experiments on Earth. That discovery provides direct evidence that the laws of physics apply across the distant reaches of the universe and from the earliest moments in time.

Atomic and molecular spectroscopy remains critical to our understanding of the cosmos, including some of the most exotic realms, from the dense plasmas on the surfaces of neutron stars to the cold dusty interiors of giant molecular clouds. Recent advances in space instrumentation have opened all ranges of the electromagnetic spectrum to astrophysical spectroscopy, which in turn has dramatically increased the need for precision atomic and molecular data. To address this need, AMO physicists carry out work in laboratory spectroscopy and study collision processes involving atoms, ions, molecules, and electrons, which often give rise to spectral lines. Theoretical calculations, benchmarked by laboratory experiments, provide essential input to the models used in the interpretation of astronomical observations.

The combination of astrophysical spectroscopy and precision atomic physics is proving to be an important testing ground for fundamental physics. As mentioned above, the spectra of distant quasars can be used to study the potential variation of fundamental constants of nature. Precision spectroscopy also played a role in the recent discovery that the expansion of the universe is accelerating rather than decelerating, as had been expected. To explain this effect, physicists have posited the existence of a "dark energy"—the nature of which is not at all understood. The very existence of dark energy poses a challenge to the Standard Model of particle physics.

Gravity, the least understood of the four fundamental forces in nature, can be investigated with astrophysical observations, specifically in its strong-field regime, where it dramatically affects the nature of space and time. For instance, looking at the exotic environment of a massive black hole or a neutron star and using the spectra of highly ionized atoms of iron and other abundant elements as precision clocks can test the predictions of Einstein's theory of general relativity in quantitative detail (see Figure 2-8).

However, the particular astrophysical conditions encountered in the vicinity of a black hole or neutron star can dramatically affect the spectra of atoms there. These environments are characterized by intense radiation fields, in which the energy density in the form of radiation can vastly exceed the thermal energy density of the gas. Atomic excitations are dominated by photoexcitation of the atoms and recombination cascades following photoionization. On the surfaces of neutron stars, gas

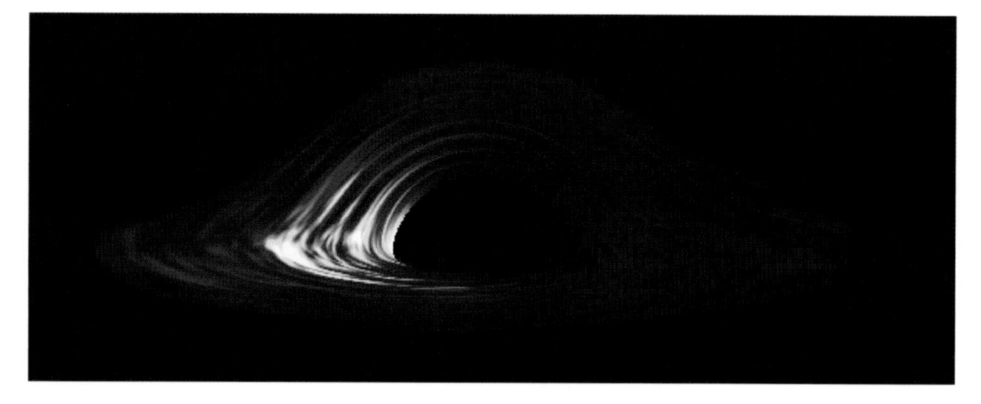

FIGURE 2-8 A simulation of the intensity pattern expected for the light emitted by a disk of material orbiting a black hole. The emission is asymmetric because the light emitted by atoms moving toward the observer is boosted in intensity, while the light from atoms moving away is diminished. The frequencies of spectral lines are similarly distorted by the intense gravitational field of the black hole. The figure also shows a strong distortion in the apparent shape of the disk—that is, the back of the disk looks like it is lifted up—due to the strong bending of light close to the black hole. SOURCE: C. Reynolds, University of Maryland.

is subjected to intense magnetic and electric fields that radically alter the atomic structure and, therefore, the spectra. Understanding such processes is a challenge for modern atomic physics. The challenges are being addressed with laboratory astrophysics experiments that use high-intensity lasers and/or particle accelerators to simulate such high-energy-density conditions. While the understanding of atomic physics under these conditions is still in its infancy, laser experiments can produce intense field regimes for short durations, as described in Chapter 4, which may be useful in this context.

The study of the early universe can yield clues to the fundamental laws of physics. Looking back in time, in the very earliest moments after the big bang the universe was compact and extremely hot. Similar conditions can be explored at accelerator facilities here on Earth using ultrarelativistic heavy ion collisions. A little later in the evolution of the cosmos, in the somewhat cooler instant of primordial nucleosynthesis, the positively charged nuclei of several chemical elements were produced—including those of hydrogen and helium together with small amounts of deuterium (HD) and lithium. The gas in the cosmos at these high temperatures was fully ionized and opaque so that it did not produce the spectral features that might be observed today.

As the universe expanded further and cooled, electrons and nuclei collided and recombined, producing neutral elements. As the universe cooled even further, molecular hydrogen and its isotope HD formed. Molecules can efficiently lose energy and allow a gas to cool under gravitational collapse, the first step in star formation. The study of molecular hydrogen formation in the early universe is an active area of study in AMO physics.

Beyond these earliest moments in the history of the cosmos, the universe only grew in complexity. Driven by the discovery of extrasolar planets and the exploration of the various moons circling planets in our solar system, astrophysicists and atomic and molecular spectroscopists are working to understand better the growth of complexity in the universe. The challenge to elucidate the evolutionary path from simple organic molecules—discovered in the interstellar medium and circumstellar shells—to simple proteins and ultimately to chemistries that give rise to life forms is exciting and compelling science. Spectroscopy is the tool with which the search for life in other planetary systems can be conducted, and AMO physics provides the backbone that supports this work.

There also remains a great deal to learn about the cosmic environment closer to home. For example, new tools enabled by AMO physics are improving our understanding of the comets in our solar system (see Box 2-4). Studying comets is expected to supply critical new information on the formation and transport of organic molecules in the solar system, as demonstrated by NASA's celebrated comet sample return mission, Stardust.

BOX 2-4
X Rays from Comets

In 1995, x-ray emission from a comet was first observed. As comets were thought to be visible because they reflect solar radiation from the cometary atmosphere, the source of the x rays was at first a mystery. Through the efforts of both atomic and molecular physicists and astronomers, the mechanism for x-ray production in comets has been identified as a "charge transfer" between highly ionized atoms in the solar wind[1] and the neutral molecules in the comet's atmosphere. Charge transfer occurs when an ion, such as seven-times-ionized oxygen, collides with a water molecule in the cometary atmosphere and captures an electron from the water, creating a highly excited state of six-times-ionized oxygen. This excited state decays by radiating light in the x-ray region of the spectrum (see Figure 2-4-1). Once this mechanism was postulated, particular x-ray spectral features could be associated with electron capture by a variety of different atomic ions in a range of ionization states. With increasing spectral resolution of x-ray telescopes in the future, coupled with further detailed calculations and measurements of charge transfer by atomic physicists, cometary x rays can be used as detailed diagnostics of the solar wind composition.

[1]The solar wind is a dilute plasma composed of protons, electrons, and a variety of atomic ions in many different ionization states. The solar wind emanates from the sun and circulates through the solar system. Earth's magnetic field largely protects it from the direct impact of the solar wind.

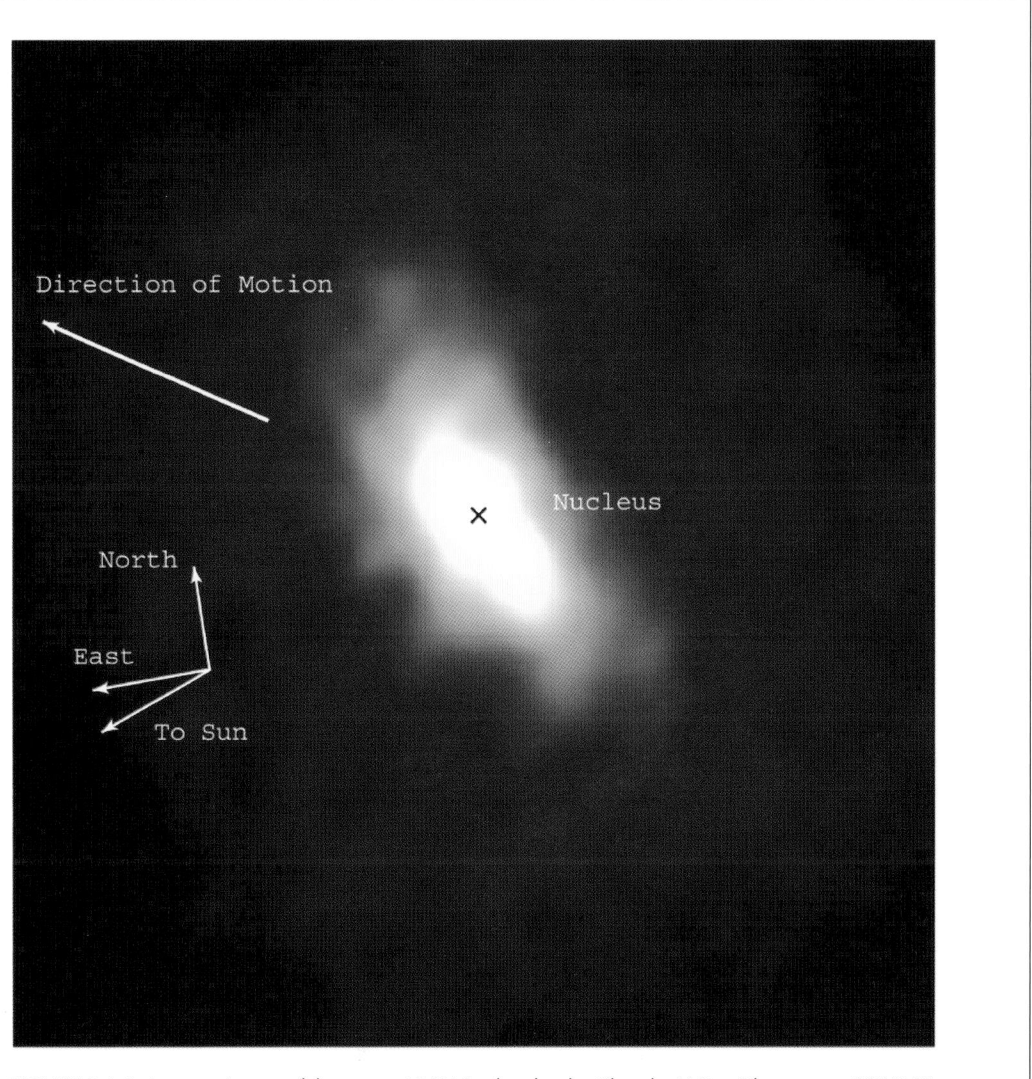

FIGURE 2-4-1 An x-ray image of the comet LINEAR taken by the Chandra X-Ray Observatory. SOURCE: C. Lisse, S. Wolk, Chandra X-Ray Observatory, National Aeronautics and Space Administration.

AMO THEORY AND COMPUTATION CONNECTIONS TO ASTROPHYSICS AND ELEMENTARY PARTICLE PHYSICS

Hand in hand with the advances in experiments, there is a continuing advance in atomic theory with applications to astrophysics and elementary particle physics. Although the Hamiltonian describing the detailed interactions of electrons in atoms and molecules is known, the calculation of the structure and spectroscopy of many-electron systems to high accuracy continues to be a challenging problem. Atomic and molecular theorists have developed powerful numerical methods and extensive computer codes to calculate energy levels, wavefunctions, and spectral line strengths. For instance, the precise cesium parity violation experiment described above would not have been so valuable without equally precise calculations of the cesium wavefunctions needed to evaluate the signal expected from particle theory. Likewise, the experiments searching for an atomic or molecular EDM rely on the atomic codes to predict the expected size of T-violating effects in atoms and molecules. Similar codes are now being used to select atomic transitions to use in the search for a time change in α. In astrophysics many spectral lines are observed from species that are difficult to produce in the laboratory. Atomic and molecular theory provides critically important input to the models used in the interpretation of these astronomical observations.

3

Toward Absolute Zero

When breakthrough science happens it defines a new frontier. AMO science is camped on one of the most exotic frontiers in science—the push toward ever-lower temperatures. When this report was written, the record low temperature stood at about a billionth of a degree above absolute zero. This is the coldest temperature in the universe (see Figure 3-1). By contrast, intergalactic space is a relatively hot 2.7 K above absolute zero owing to the cosmic microwave background. The frontier of this research is at the intersection of AMO with other fields, particularly condensed matter physics, low-temperature physics, plasma physics, and even theoretical nuclear physics. The interaction of researchers in these different fields is leading to exciting new physics, promising a decade of rapid advancement in these areas of science and blurring the line between research fields. In the last decade, six physicists have won Nobel prizes for their work at the frontier of the ultracold, and we are just beginning.

THE PROMISE OF ULTRACOLD SCIENCE

The first step in almost any ultracold gas experiment consists in cooling a gas of atoms to "degeneracy." This means that it is so cold that the de Broglie wave of one atom starts to overlap with that of its nearest neighbor (see Box 3-1). When the atoms in question are bosons, the result is a Bose-Einstein condensate (BEC). Below a certain critical temperature, systems of trapped bosonic atoms become Bose-Einstein condensed superfluids, with a macroscopic fraction of the particles occupying the lowest-energy "wave" in the box in which the atoms are kept. In 1995 the first Bose-Einstein condensed gases of bosonic alkali atoms were produced (see Figure 3-2). BECs are by now routinely studied worldwide in systems of millions or more atoms, down to temperatures some billionths of a degree above absolute zero.

The striking characteristic of a superfluid is its ability to flow without even the

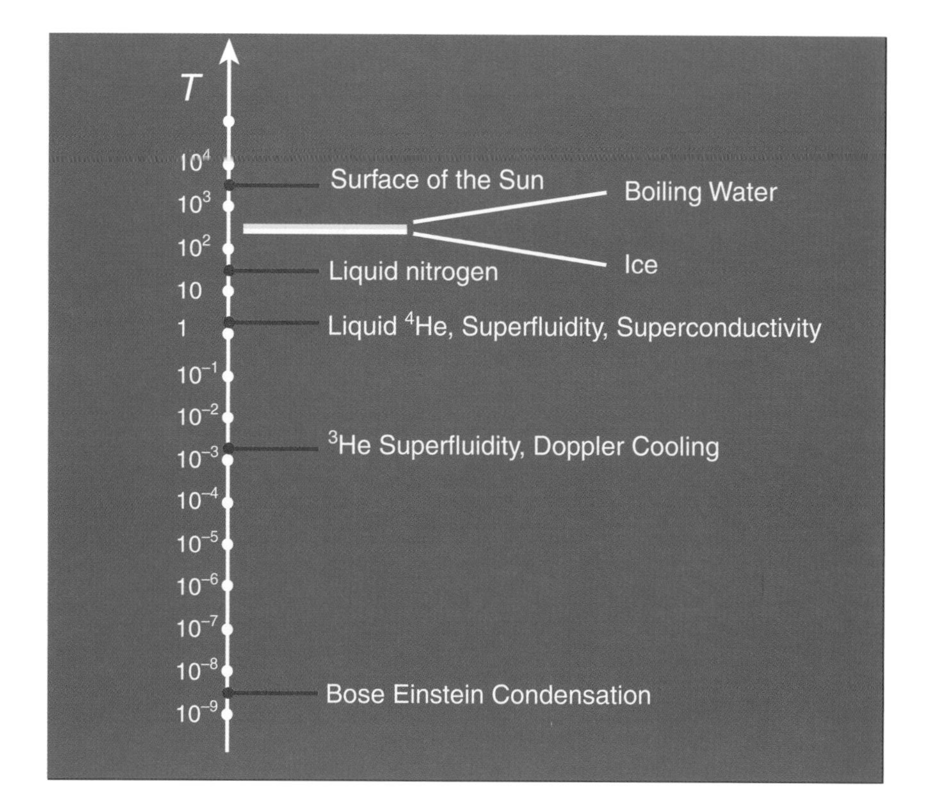

FIGURE 3-1 Temperatures of some familiar objects, on a scale of powers of 10.

BOX 3-1
de Broglie Waves

It has been known since the work of Max Planck and Albert Einstein that one must sometimes think of light as consisting of (massless) particles, now called photons. This idea complements the more common notion that light is a wave. Likewise, quantum mechanics teaches us that atoms, as well as all particles, also possess wavelike properties. This wave-particle duality of both light and atoms is the cornerstone of quantum mechanics, yet it remains one of its most unsettling aspects.

The quantum-mechanical wavelength of massive particles in thermal equilibrium is inversely proportional to the square root of their temperature. It is exceedingly small at normal (room) temperatures, a small fraction of a nanometer or so. As atoms get colder, though, their wavelength becomes longer—for the temperatures we will discuss here, these wavelengths can be hundreds of micrometers. Another way to say this is that at ever-lower temperatures, quantum mechanics becomes progressively more dominant.

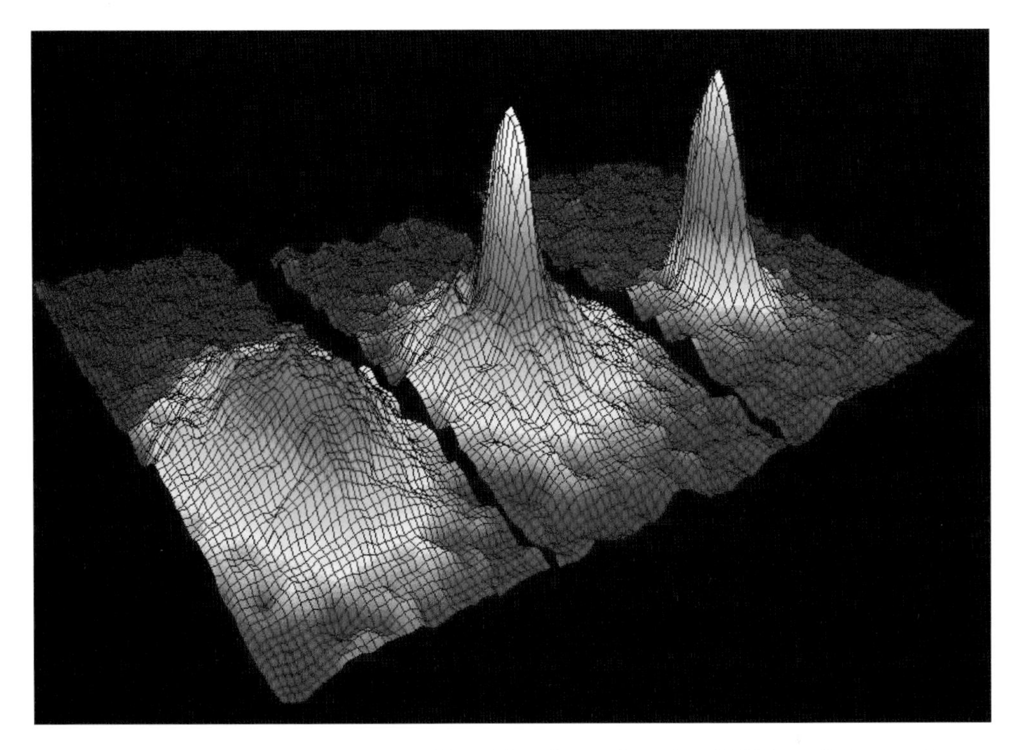

FIGURE 3-2 The original demonstration of Bose-Einstein condensation (BEC) in a dilute gas of rubidium atoms. These false-color images show the velocity distribution of a cloud of cold atoms near and below the BEC critical temperature. *Left:* Above the transition temperature, the atomic velocities show that this is an ordinary cold gas at an extraordinary temperature of 400 billionths of a degree above absolute zero (400 nK) . *Center:* Further cooling to 200 nK leads to the onset of BEC, shown as a clumping of atoms near zero velocity. *Right:* Still more cooling to 50 nK shows nearly all of the remaining gas in the BEC phase. SOURCE: E. Cornell, C. Wieman, JILA.

slightest resistance, to—as if by magic—overcome friction. Superfluidity is very much analogous to superconductivity, the ability of a metal to conduct electricity without any loss. Indeed, in a very real sense superconductivity *is* superfluidity, with the electrical current being carried along through the superconducting metal as a superfluid of electrons. A superfluid is to a regular fluid what a superconductor is to a regular electrical conductor. Until the mid-1990s the only laboratory superfluids known were liquids of both helium isotopes, ^{4}He and the rarer ^{3}He.

In principle, a superfluid gas flowing around a closed loop can keep flowing (or "persist") forever. At a conceptual level, persistent superflow around a loop can be understood as analogous to a twisted loop of ribbon. Imagine taking a length of ribbon, putting a twist in it, and then bringing the two ends of the ribbon together and permanently gluing them. The twist in the loop of ribbon represents the flow

BOX 3-2
Fermions and Bosons

According to the basics of quantum theory, any two electrons are identical. In fact, any two atoms, so long as they are the same kind of atom, are also identical. This simple statement has some extremely far-reaching consequences. If we exchange any two identical particles, the system will act in precisely one of two ways, depending on whether the particles are fermions or bosons. Composite particles composed of fermions may be either fermions or bosons, depending on the number of fermionic constituents. Particles composed of an even number of fermions are themselves bosons; those composed of an odd number of fermions are themselves fermions.

Fermions are constrained by the Pauli exclusion principle, which states that only one fermion can occupy a given quantum state at a time. Bosons, on the other hand, are not constrained by the Pauli exclusion principle. At low temperatures, bosons can therefore collect into the same energy state. In fact, the more bosons that are in a state, the more likely it is that still more will join, resulting in Bose-Einstein condensation.

of the superfluid. The persistence of the flow is just a consequence of the difficulty one would have in removing the twist from the loop without cutting the loop. One can crumple the loop, one can stretch it, but the twist is trapped in the loop by topology. "Cutting the loop" in this analogy means destroying the BEC locally. But at sufficiently low temperatures, the gas wants to remain in a condensed state. So while the unavoidable effects of friction along the flow of the fluid try to stop the flow, until there is a concentrated source of energy to destroy the condensate, it will flow forever.

Atoms come in two kinds: "social" bosons and "loner" fermions (see Box 3-2). The fermions are as important as the bosons that we have talked about so far. Indeed, there would be no chemical elements and no life as we know it if fermions were not the way they are. Aside from fermionic atoms, a fermionic particle of particular importance is the electron. Most properties of chemical elements are determined by electrons, and many properties of solids are determined by them as well. This includes not just their electrical conductivity (or superconductivity) but also the thermal conductivity and magnetic properties of metals. Atomic physicists in the last few years have extended the study of ultracold, degenerate gases to include fermionic atoms.

CONDENSED MATTER PHYSICS IN DILUTE ATOMIC SYSTEMS

In ordinary liquid or solid matter, atoms are tightly packed and pressed against each other, typically separated by a fraction of a nanometer. Interactions between atoms are complicated and often not exactly known. In contrast, in dilute gases the density is a billion times lower: The atoms are separated by thousands of nanometers, a regime where the interactions are well known and characterized. In addition, instead of interacting with a large number of neighbors, the interactions between atoms tend to be largely pairwise only, considerably simplifying the theoretical description of these systems as well as their control in the laboratory. At such distances, however, the normal short-range effects of quantum mechanics, so critical to the physics of solids, are largely suppressed.

One important application of ultracold atom physics is to develop scale models that can tell us about condensed matter systems. To construct such systems as we scale up the distance between atoms, we have to scale the temperature down, way down, so as to increase their de Broglie wavelength. This renders quantum mechanics once more dominant. Indeed, temperatures a billion times lower than usual are required to mimic, in a gas, the physics of a solid. Fortunately, nano- or even picokelvin temperatures can now be achieved in many laboratories using a combination of cooling techniques.

Tuning the Interactions Between Atoms

A very powerful experimental tool available in cold-atom studies is the ability to adjust the strength, and even the nature, of the interaction between nearby atoms. This is accomplished simply by adjusting the ambient magnetic field near a so-called Feshbach resonance (see Box 3-3). As the magnetic field approaches the resonant value from one side, the interactions are attractive and become extremely strong. On the other side of the resonance, the interactions are effectively repulsive. One can literally dial in the interaction one wants—whether strong, weak, or zero, whether attractive or repulsive—and one can even change it as an experiment is in

BOX 3-3
Feshbach Resonances

During a collision, two atoms can stick together for a brief time and form a short-lived molecule. If the molecular state has a different energy than the colliding atoms, this time is very short. If the magnetic field produced by the electrons in the molecule differs from that of the atoms, one can use an external magnetic field to vary the energy difference between the atomic and molecular states. Near the field where the energies match, the so-called Feshbach resonance, the atoms can stick together for a long time. On one side of the resonance, the effective interaction between the colliding atoms is strongly repulsive; on the other side it is strongly attractive. Right at the resonant field, atoms can form molecules without release of any heat. This has opened up the possibility of ultracold chemistry and coherent conversion of atoms into molecules. Recently, such ultracold association processes have been extended to three and four atoms.

progress. This extremely powerful tool, without a counterpart in condensed matter physics, has led to remarkable developments such as the discovery of high-temperature superfluidity in a fermionic gas and the creation of molecular condensates. It also offers extraordinary future promise—for instance, in quantum information science, the topic of Chapter 7.

When the effective interaction between two ultracold fermionic atoms is repulsive, then there is a closely lying bound state of a bosonic diatomic molecule. When the effective interaction is weakly attractive, on the other hand, and for low enough temperatures, the fermions favor arranging themselves into weakly bound pairs of atoms of opposite momentums called Cooper pairs. This pairing mechanism was discovered by Bardeen, Cooper, and Schrieffer (BCS) to lie at the origin of superconductivity.

As noted, it is possible simply by varying the ambient magnetic field to tune a system from the "repulsive" side of the Feshbach resonance, where the atoms can combine into stable molecules and form a molecular BEC, to the "attractive" side of the resonance, where the situation is as in superconductivity. The difference is that the formation of Cooper pairs is now between fermionic atoms instead of between electrons. In this way it is possible to transform a molecular condensate into a superfluid of paired fermionic atoms. It had never been possible in the past to study this BEC-BCS crossover in the laboratory, but the availability of ultracold fermionic atomic gases now makes such studies possible.

High-temperature superconductivity, a discovery that made the newspaper headlines in the 1980s, is not yet fully understood and remains a daunting challenge for theoretical condensed matter physics. In ultracold fermionic gases, a new form of high-temperature superfluidity was discovered in a breathtaking series of experiments performed in 2004 and 2005. These observations are now the starting point for an investigation of different mechanisms for pairing fermions and superfluidity, and there is much hope that this work will contribute significantly to understanding high-temperature superconductivity in solids. A goal of such studies is to develop systems that have the more realistic features of high-temperature laboratory superconductors.

Optical Lattices

Many experiments with cold gases are performed with the samples confined in bowllike traps. Electrons and atoms in solids, on the other hand, live not in a bowl but in a corrugated, undulating, egg-carton-like potential that arises from the ions of the solid being arrayed in a crystalline lattice. An increasing number of ultracold atom experiments are being done in periodic potentials as well, with the periodicity being imposed experimentally by means of an optical lattice.

Using optical lattices—formed by counterpropagating lasers producing standing electromagnetic waves—one can control to an unprecedented degree the environment in which the atoms sit. For example, one can produce lattices in one, two, and three dimensions with a wide variety of lattice spacings and structures in which neutral atoms can be trapped (see Figure 3-3). When the depth of the corrugation is relatively shallow, the atoms moving through it pick up the qualities of electrons in a good conductor and can move almost freely within the crystal.

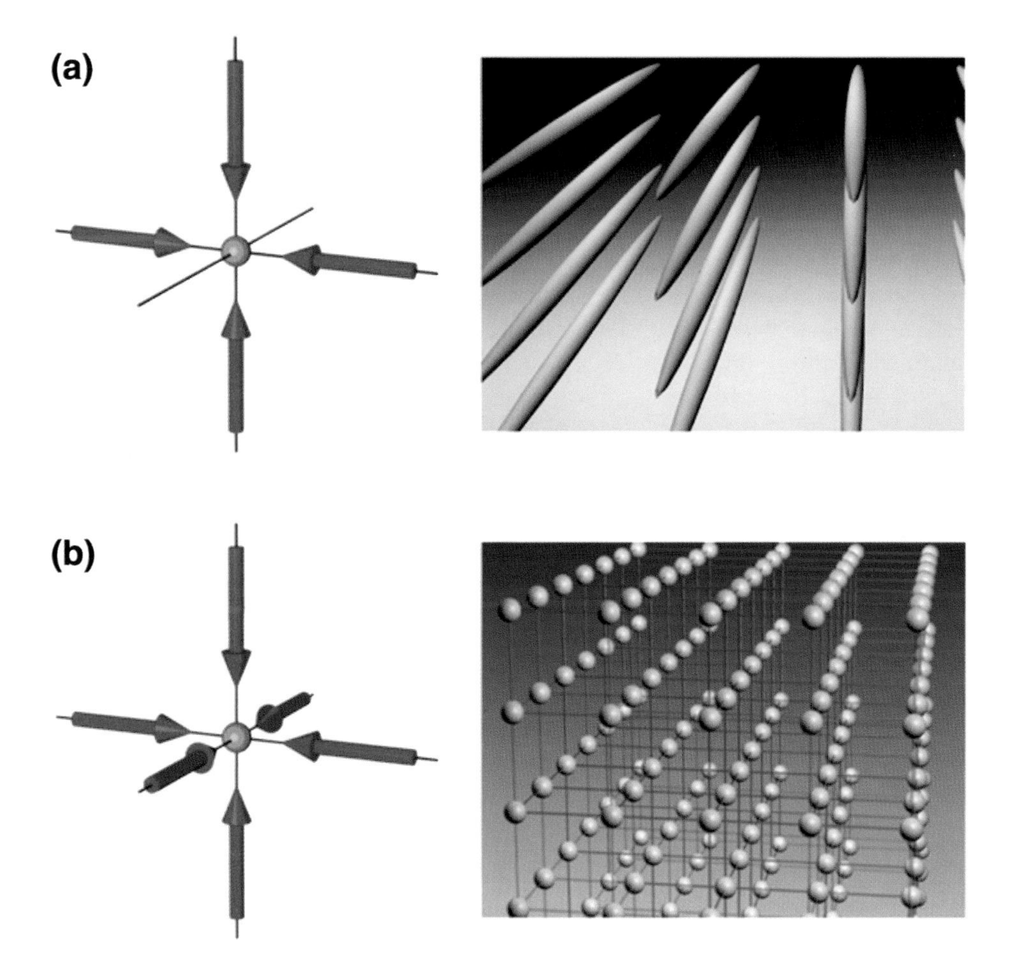

FIGURE 3-3 Optical lattices are periodic potentials formed by the intersection of several laser beams in ultra-high-vacuum chambers. The left-hand panels show laser configurations forming (a) two- and (b) three-dimensional potentials. The right-hand panels are schematics of the corresponding arrays of neutral atoms that can be trapped by these potentials. SOURCE: I. Bloch, Johannes Gutenberg University, Mainz, Germany.

As the lattice wells become deeper, more exotic condensed matter behavior kicks in. For instance, one can convert from a superconducting state, where the atoms are delocalized over the whole lattice much like electrons in a superconductor, to a so-called Mott insulator state. In the latter state, each lattice site is host to a precise number of atoms, and there is no possible transport of atoms from one site to the next (see Figure 3-4).

Optical lattices give us the ability to simulate a vast range of conditions expected in condensed matter systems, such as high-temperature superconductivity in layered oxides and numerous other exotic states of quantum matter. Indeed, in such situations the trapped atomic systems can be thought of as a form of quantum simulator (see discussion in Chapter 7). Using trapped atoms holds great promise; using cold atom systems directly to simulate the behavior of various physical systems is already a possibility with optical lattices. For example, the Hubbard model, a collection of particles localized on a lattice and interacting with their nearest neighbors, is a key system studied in condensed matter physics. By constructing in the laboratory an analog system of fermionic or bosonic atoms trapped in an optical lattice, it is possible to measure the properties of the Hubbard model over wide ranges of dimensionality, lattice filling, interaction strength, temperature, and more. In addition, one can also exploit the atomic spin to achieve an additional class of simulations: In both fermionic and bosonic systems, the atomic spin is an important feature that influences both microscopic, single-atom behavior and collective behavior, like superfluidity. Recent experiments provide evidence for a new

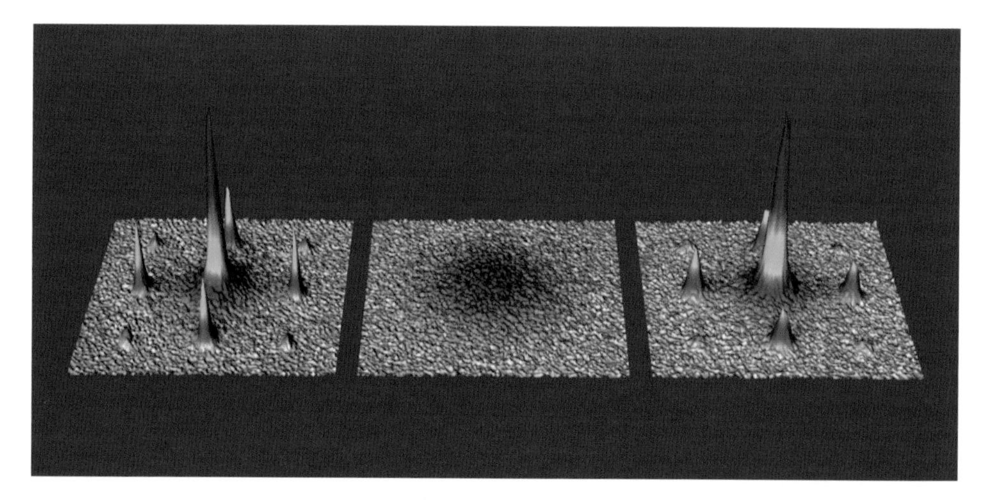

FIGURE 3-4 Momentum distribution of ultracold atoms in an optical lattice, illustrating the transition from a superfluid to a Mott insulator state and back. SOURCE: I. Bloch, Johannes Gutenberg University, Mainz, Germany.

state of matter, a supersolid, in which solid helium exhibits superfluidity at low temperatures. This remarkable coexistence of ordering of atomic states underlying superfluidity and crystalline ordering in the absence of an external periodic potential was generally unexpected. In optical lattices, which have such a potential, the two kinds of order have been observed to coexist. Studying the behavior of bosonic atoms in such lattices as the potential is turned down to zero might give us insight into viable scenarios for superfluidity in solid helium. Finally, the extension of optical lattices to molecular systems increases the size and utility of the ultracold atom "toolbox" even more, a point to which we return shortly.

Vortices

Another tool in the simulated solid toolbox is the vortex. When a superfluid is rotated, the rotation is concentrated into tiny tornado-like quantized units called vortices (see Figure 3-5). At fast rotations, which can be generated simply by "stirring up" the bowl in which the condensate is confined, the superfluid is pierced by hundreds of these tiny twisters, which organize themselves into regular arrays, rows, and columns. A whole new class of exotic states of matter is predicted to occur at high rotational speeds, when the number of vortices approaches the number of atoms in the condensate. This system is mathematically closely related to the quantum Hall effect of electrons in two-dimensional semiconductors, which has led to an unprecedented accuracy in the measurement of electrical resistances. Such exotic states are now close to being produced in Bose systems and are expected in Fermi systems as well.

The overlap between cold-atom physics and condensed matter physics described in the previous paragraphs will create stunning opportunities for advances in experiment and theory over the next decade.

MOLECULES AND CHEMISTRY

While some of the most fascinating science in the past decade has emerged from the study of ultracold atoms, the coming decade may well be the decade of cold molecules. In the last 2 or 3 years, there have been a number of breakthroughs in the efforts to extend the techniques of cold atoms to the nascent field of cold molecules.

A molecule in the simplest case with only two atoms is much like a tiny barbell with one atom at each end. The molecule can vibrate, with the barbell becoming alternately longer and shorter, and it can rotate, whirling end over end like a tiny baton. Chemical reactions take place during collisions, when molecules hit each other randomly, sometimes end to end, sometimes in the middle. A key means of

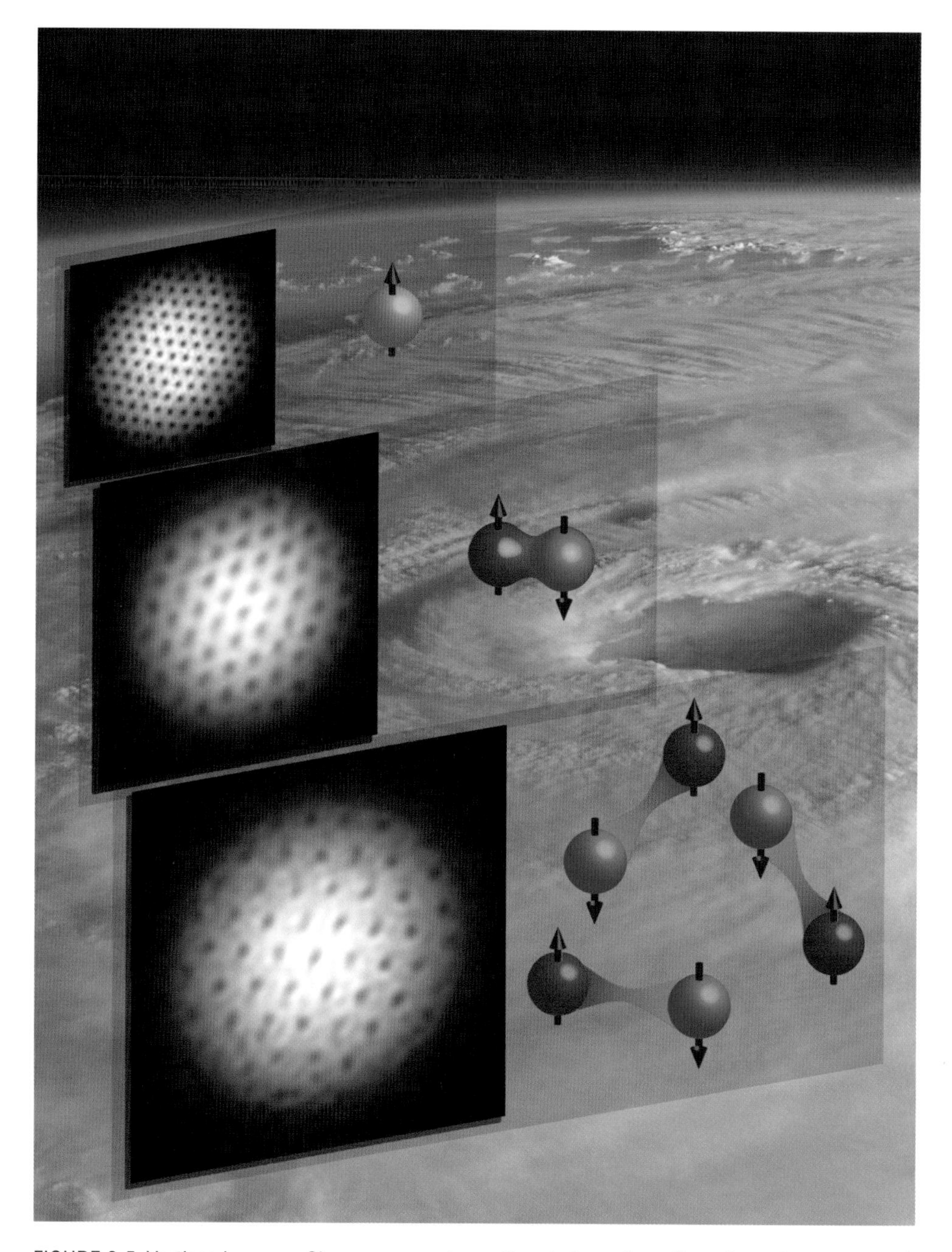

FIGURE 3-5 Vortices in gases. Shown are a vortex pattern in bosonic sodium atoms (green cartoon) in a magnetic trap; vortices in tightly bound lithium molecules (red-blue cartoon); and a vortex lattice in loosely bound fermion pairs on the Bardeen-Cooper-Schrieffer (BCS) side of a Feshbach resonance. The background shows a classical vortex (Hurricane Isabel in summer 2003). SOURCE: W. Ketterle, Massachusetts Institute of Technology.

controlling these collisions, and thereby controlling molecular chemistry, is to cool the molecules to low temperatures.

Why is control of chemistry so dependent on the temperature of the colliding molecules? The point is that the random molecular orientations that are normally present create a lack of control over the collision processes that lead molecules to form new compounds. In many molecules, one end is electrically positive and the other is electrically negative. The electrical charge separation in a molecule means that pairs of molecules interact with each other over relatively large distances. At low temperature, simply applying an electric field can make all the molecules line up. This allows the detailed study of chemical reactions with exquisite control, allowing scientists to tease out the fundamental nature of molecular interactions. This fundamental understanding will likely affect numerous areas of science and technology 20 years from now.

In the last half-decade, there have arisen two rather distinct approaches for creating cold molecules. The first proceeds by first creating molecules using conventional "hot" chemistry, then cooling them down via collisions with cold noble gases. The second approach is to first create ultracold atoms and then induce them to stick together, to react and form molecules, in such a way as to preserve the initial low temperatures. These two basic approaches have their own relative strengths and weaknesses. The first approach is more general: It can work for many different kinds of molecules. The second approach is more spectacular: The ultimate temperature reached by recombination can be nearly as low as the temperature of the coldest atomic BECs and recently led to the formation of a molecular BEC. It is hard to predict which will be the more widely practiced method 10 years hence. Most likely both will still be in play, because their strengths are complementary and the choice will depend on the ultimate application.

The concept of superfluidity is discussed earlier in this report, with the persistence of the flow explained as analogous to the difficulty in removing a twist from a ribbon loop without cutting the ribbon. A fascinating possibility with molecular condensates is that the internal degrees of freedom of the molecules may change the rules that govern taking the twist out of a loop. The metaphorical ribbon that describes the superfluid flow in a molecular gas is embedded in a higher-dimensional space than an ordinary twisted ribbon. Under certain circumstances, one may be able to bend and stretch this higher dimensional ribbon to remove the twist without tearing the ribbon! Some higher dimensional "ribbon worlds" allow for this untwisting, and some do not. The consequences of this seemingly unworldly but very real distinction might be seen in the presence or absence of friction in a BEC of molecules.

Once the formation of ultracold molecules has become somewhat routine, it will also become possible to start exploring chemistry near absolute zero. For example, it is possible that in the future Feshbach resonances could be exploited to enhance the rate of specific chemical reactions by orders of magnitude and to achieve an exquisite degree of selectivity between competing product channels.

Chapter 7 contains an extended discussion of quantum information and quantum computing. In the context of a discussion of future research in cold molecules, it is worth noting that the key technical challenge in quantum computing is to develop a system of quantum bits that interact with one another so as to form gates but that do not interact with their environment, so as to avoid decoherence. Trapped neutral atoms present one possible AMO realization of this, as do trapped ions (see Chapter 7). Neutral atoms are well isolated from the environment, and do not easily interact with one another; on the other hand, ions interact easily with one another, but their interactions are too long range, and they are more sensitive to their environment. Ultracold molecules in an optical lattice may offer an optimal compromise here.

ATOM OPTICS

The de Broglie wavelength of atoms is inversely proportional to the square root of their temperature. It is exceedingly small at normal (room) temperatures, about the size of the atom itself. This is why it is normally difficult—although not impossible—to observe the wave nature of atoms. At the extreme low temperatures that can now be achieved by laser cooling and other cooling techniques, however, the wave nature of atoms and molecules completely dominates their behavior. The analogy between atomic and optical waves can then be exploited in the emerging field of atom optics. For example, it is possible to build a matter-wave analog of an optical interferometer, a device that holds great promise for both fundamental science and applications (Box 3-4). Chapter 2 discusses how these interferometers are beginning to find applications in precision measurements. See Figure 3-6 for an example of a matter-wave interferogram.

Before the realization of atomic Bose-Einstein condensation in 1995, the situation in atom interferometry had been similar to the situation in optics before the invention of the laser, and the spatial and temporal coherence of sources required to produce interference fringes was largely obtained by filtering techniques, much as had been the case in traditional optics. This approach, which does not necessitate the use of ultracold atomic or molecular beams, let alone quantum-degenerate ones, is still used in many groundbreaking experiments on matter-wave interferometry.

While it is relatively easy to observe quantum interferences with small objects such as electrons and atoms, observations become harder for larger and larger molecules, which interact more strongly with their environment. Molecular-interference

BOX 3-4
Atom Interferometers

In a conventional optical interferometer, a light beam is split into two or more paths that are recombined later on a detector. If the beams from the different paths are in phase, their fields add up and the detector measures bright light. If they are out of phase, on the other hand, the fields cancel each other and the signal at the detector indicates darkness. Similarly, in an atom interferometer, as in Figure 3-4-1, one can observe the coherent addition or cancellation of matter waves. By carefully monitoring the variations of "bright" and "dark" periods, one can determine with exquisite precision the physical processes that influenced the dynamics of the atoms in the two paths of the interferometer. For instance, atom interferometers have been used to carry out tests of Einstein's equivalence principle at the atomic level and are being developed into rotation and inertial sensors with unprecedented sensitivity and accuracy.

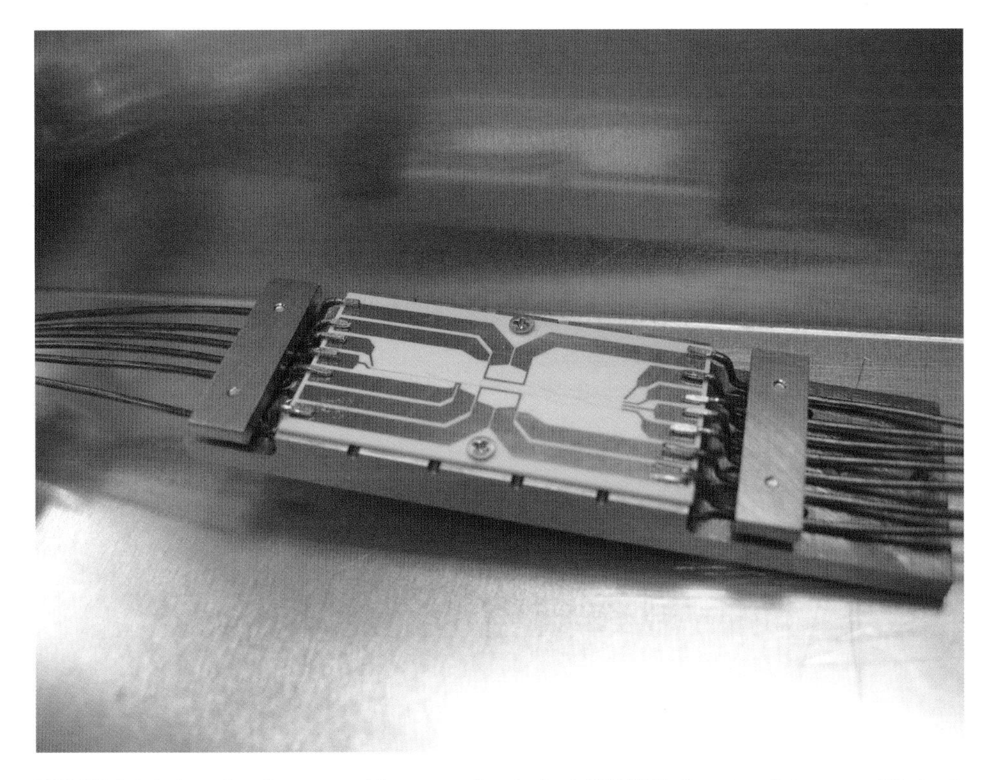

FIGURE 3-4-1 Atom interferometer chip mounted and wired. SOURCE: Quantum Sciences and Technology Group, Jet Propulsion Laboratory, National Aeronautics and Space Administration.

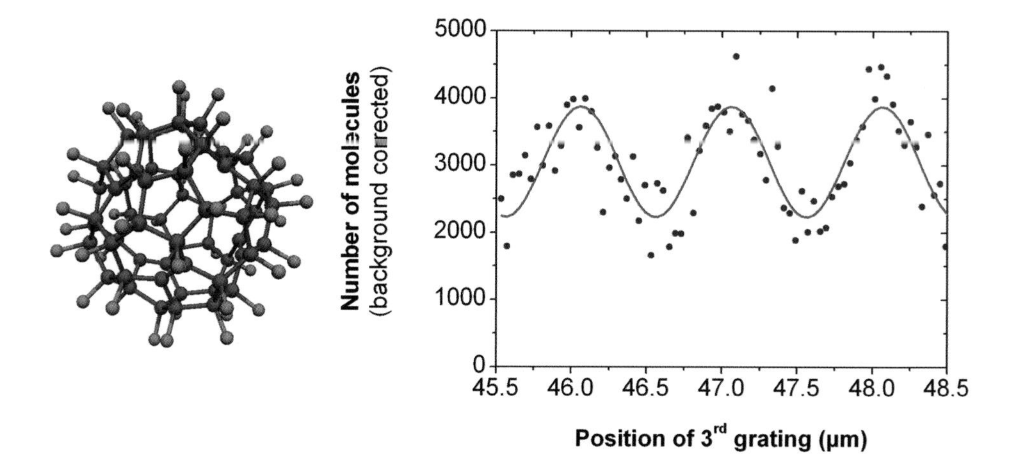

FIGURE 3-6 Interferogram (*right*) of $C_{60}F_{48}$ (*left*). This molecule currently holds the record in complexity (108 atoms) and mass (1,632 amu) for matter-wave interferometry. This molecule exists in two isomers of different symmetry. Both were present in the same experiment. SOURCE: M. Arndt, L. Hackermüller, and A. Zeilinger, Faculty of Physics, University of Vienna, Austria.

experiments are far from easy, but recent work shows that large, complex molecules can indeed interfere and reveal their quantum nature. Such experiments have now been carried out with molecules containing 100 or more atoms bound in a single interfering object (see Figure 3-6). Especially in the context of quantum information science, understanding the mechanisms of decoherence is a central issue. In matter-wave interferometry, decoherence can be traced back to the fact that the molecule exchanges information with its environment via thermal photons, the radiation emitted by any object with a temperature that is not absolute zero. Specifically, each photon emitted by the molecule transfers information about the molecule position, and indeed decoherence due to heat radiation is of particular relevance for macroscopic objects.

In a BEC near absolute zero, practically all atoms are in the same quantum state. This situation is similar to that in a laser, where a vast number of photons have precisely the same wavelength, direction, and phase. Extracting an atomic beam from the BEC makes it possible to realize the matter-wave analog of a laser beam (see Figure 3-7).

Nonlinear Atom Optics

Many laser applications rely on the ability to mix photons of different wavelengths and/or directions of propagation using techniques of nonlinear optics. For

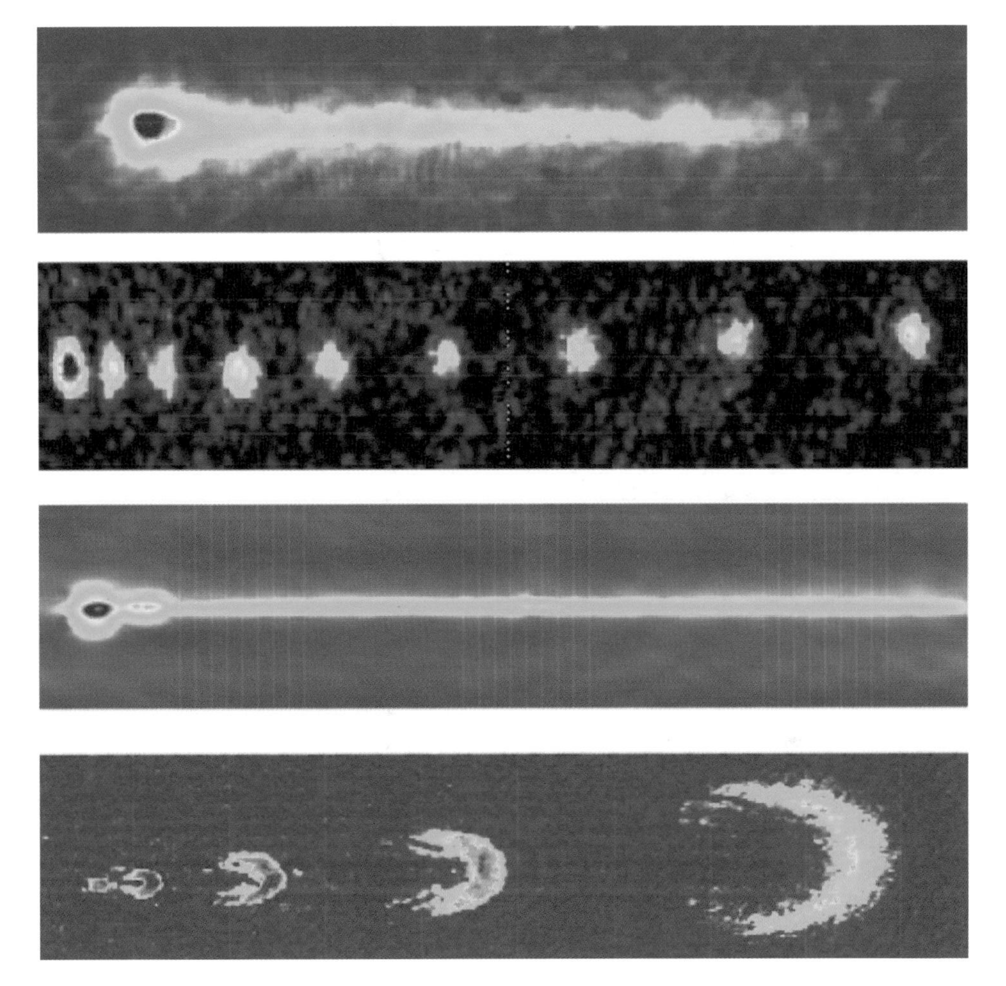

FIGURE 3-7 Picture gallery of coherent atom beams, so-called atom lasers, extracted from a BEC, the red dot at the left of each figure. The beams are shown both pulsed and continuous. SOURCE: W. Ketterle, Massachusetts Institute of Technology; I. Bloch, Johannes Gutenberg University, Mainz, Germany, and T.W. Hänsch, Max-Planck-Institut für Quantenoptik, Garching, Germany; M. Kasevich, Yale University; and W.D. Phillips, National Institute of Standards and Technology.

instance, it is possible to annihilate two photons of one frequency to produce a new photon at double the frequency (second-harmonic generation) or to mix three optical fields to produce a fourth one (four-wave mixing). There is also an atom-optic analog to nonlinear optics. Four-wave mixing in matter waves—the coherent generation of a new matter wave out of three incident waves—and the launching of solitons (waves that can propagate long distances without changing their shape) soon followed the first experiments in BEC. The mechanisms by which two atoms combine into a diatomic molecule can be thought of as analogous to second-harmonic generation in optics, except that instead of annihilating two photons to create a new one with double the frequency, we now have the annihilation of two atoms and the creation of one molecule (with double the mass). In addition, it has proven possible to mix optical and matter waves.

The Pauli exclusion principle forbids putting two identical fermions in the same quantum state and hence building a fermionic atom laser. Nonetheless, fermionic atoms may offer a viable alternative to bosons for interferometric applications that require the detection of tiny path differences in the arms of the interferometer. This is somewhat analogous to the situation in white-light interferometry, which has proven exceedingly useful in many conventional optics applications. An advantage of quantum-degenerate fermions in this context is that they do not suffer the kind of collisions that bosons are subject to. These collisions can lead to decoherence and the destruction or random shift of an interference pattern. The absence of collisions between ultracold fermions is another direct consequence of the Pauli exclusion principle.

Integrated Atom Optics

In electronics and in conventional optics, the miniaturization of components and their integration into networks leads to powerful tools and devices. Similarly, integrated atom optics could result in powerful devices that combine many atom optical elements into integrated quantum matter-wave circuits. Current research seeks to develop atom chips to implement atom-based devices on a small scale. On a single substrate one can, for example, incorporate conducting wires, magnetic elements, and optical components to produce miniaturized atom optical elements that confine, control, and manipulate matter waves. On that same substrate one can also incorporate atom detection and signal conversion elements. Atom chips are now being developed for a whole new range of applications, from novel sources of quantum-degenerate gases to compact atom interferometers. In combination with techniques of cavity quantum electrodynamics and with microfabricated resonators, they may also lead to the development of novel systems for quantum information technologies.

Quantum Atom Optics

Improvements in detectors will transform atom optics to quantum atom optics, the matter-wave analog of quantum optics. Many future applications of atom optics will be tied to the development of improved neutral atom detection schemes.

As experiments become more and more sophisticated, it will become increasingly important to go from simple digital photographs of the clouds and moving pictures that accurately record the density and spin patterns of the samples, to detection schemes that can address questions related to the precise quantum state of the system under study. They will include single-atom detectors using detectors based on optical resonators developed for cavity quantum electrodynamics, charge-coupled devices, or sophisticated image-processing techniques.

REACHING OUT: PLASMAS, NUCLEAR PHYSICS, AND MORE

New and unforeseen bridges are developing between ultracold physics and areas of research that might at first sight seem far removed, including systems under extreme conditions such as plasma physics, high-energy physics, and astrophysics.

Cold Plasmas

Plasmas are ubiquitous in the universe. Most of observable space is in a plasma state, so that plasma physics is the natural science that we apply to understand the phenomena of space. Clouds of charged particles, electrons, and ions surround Earth. Approximately 99 percent of the interstellar medium is composed of interstellar gas that consists partly of neutral atoms and molecules, as well as charged particles such as ions and electrons. Plasma physics and astrophysics require detailed atomic spectra and structure calculations to diagnose and predict properties of plasmas, as well as to understand the temperatures and velocities of astrophysical objects.

Ultracold atoms add a new frontier to plasma physics: the creation of plasmas colder than any ever created before (Box 3-5). Cold neutral plasmas, with equal numbers of positive and negative charges, are challenging systems for computational physics, in particular for the types of molecular dynamics calculations that are also used to simulate thermonuclear plasmas. Cold neutral plasmas can recombine into neutral atoms—and this is at the heart of the experiments looking at the formation of the antimatter version of hydrogen, antihydrogen. Another fascinating aspect of cold plasmas is that they can become so cold that instead of acting like a gas, they behave more like a liquid or even a crystalline solid. Such strongly coupled plasmas are thought to lie at the core of white dwarf stars and possibly massive Jupiter-sized planets.

Nonneutral plasmas are created from laser-cooled ions held together by strong magnetic fields. These plasmas can form beautiful "crystals," with symmetries and structures depending on the shape of the confining field (see Figure 3-8). These

BOX 3-5
Cold Plasmas

At cryogenic temperatures, trapped-ion plasmas become strongly coupled and, at sufficiently low temperatures, form crystalline states. These crystalline states have been observed and studied with laser-cooled atomic ion plasmas stored in Penning and radio-frequency traps. Strongly coupled plasmas are models of dense astrophysical matter, as well as quark-gluon plasmas produced in ultrarelativistic heavy ion collisions. Recently it was theoretically shown that laser-cooled ion plasmas in a Penning trap could be used to measure the exponential enhancement of close collisions in a strongly coupled plasma. This is exactly the same enhancement predicted over 50 years ago for the nuclear fusion rate in dense stellar interiors. Future work can provide the first precision measurements of this enhancement. This is yet another example of how measurements done on model AMO systems can advance our understanding of condensed matter systems (in this case, our understanding of nuclear reactions in dense stellar interiors).

Small, one-dimensional crystal strings of three or four ions have been used to demonstrate quantum gates and some of the basic building blocks of a quantum computer. These few ion experiments demonstrate the control that is possible with trapped ions. Long relaxation times of greater than 100 seconds have been measured on ground-state hyperfine coherences in cold plasmas of trapped ions. This combination of control and long coherence times makes cold plasmas of trapped ions a fertile ground for investigating the generation of entangled or squeezed spin states, which can be used in sub-shot-noise spectroscopy.

In recent years, a significant effort in the AMO community has been going into cooling, trapping, and controlling molecules in the gas phase. In trapped-ion plasmas, cooling proceeds quite simply by using an atomic ion species that can be laser-cooled to sympathetically cool molecular ions. In this manner crystals of molecular ions have been formed in an ultrahigh vacuum, without perturbation of the internal states for periods longer than many minutes. These translationally cold molecular ions provide a good starting point for precision spectroscopy and for the investigation of techniques to control the molecular populations.

systems probe the fundamental aspects of self-organization, with only simple electric repulsion leading to highly organized states. Further applications of cold, trapped plasmas range from the generation of intense positron beams for material science studies to the generation of squeezed spin states that can be used to improve atomic clocks.

The Bose condensation of Cooper pairs, discussed earlier in this report in the context of superconductivity in metals, is also familiar in nuclear physics, where BCS pairing explains reduced moments of inertia of heavier nuclei. In astrophysics, condensates of pi mesons and of K mesons have been studied as possible states of neutron star matter, where both the neutron and proton components undergo BCS pairing to become superfluid. Such superfluidity most likely underlies the observed glitches, or sudden rotational speedups, observed in some 30 pulsars to date.

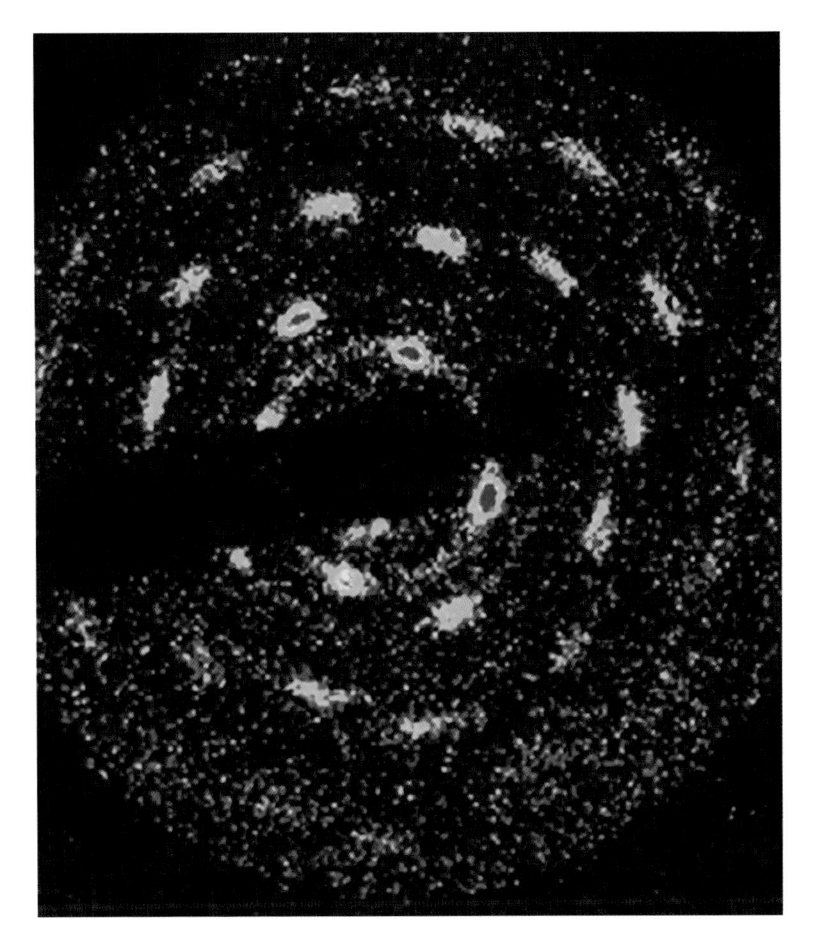

FIGURE 3-8 Time-resolved Bragg scattering is used to obtain information on the spatial correlations of trapped and laser-cooled beryllium. The pattern is consistent with Bragg scattering off a single body-centered cubic (bcc) crystal. Analysis of gated and time-averaged patterns like those shown here indicates that with several hundred thousand trapped ions, bcc crystals are always observed. SOURCE: National Institute of Standards and Technology.

A close connection has emerged between the quark-gluon plasmas formed in heavy ion accelerators such as Brookhaven National Laboratory's relativistic heavy ion collider (RHIC) facility and systems of ultracold trapped laboratory fermions near the BEC-BCS crossover (Box 3-6). Both systems are extremely strongly coupled forms of matter. As observed both at RHIC and in the AMO laboratory, such matter flows with minimal viscosity. Understanding such connections will be a fascinating challenge to both high-energy nuclear and AMO theories.

BOX 3-6
Boson Condensates at the Relativistic Heavy Ion Collider (RHIC)

Bosonic condensates formed from the pairing of spontaneously created quarks and antiquarks are fundamental features of the vacuum and the structure of elementary particles such as the nucleon; such condensates underlie the spontaneous breaking of the chiral symmetry (the symmetry between right- and left-handedness) of the strong interactions. One of the important aims of ultrarelativistic heavy ion collisions at the RHIC and, starting in 2007, at the Large Hadron Collider (LHC) is to produce chirally restored matter in the form of quark-gluon plasmas. There, one is asking the opposite of the question asked in condensed matter physics—namely, What are the properties of Bose-Einstein "de-condensed" matter?

THE SYNERGY BETWEEN EXPERIMENT AND THEORY

AMO science enjoys a very close interaction and synergy between theory and experiment, which benefit each other in numerous ways. Indeed, the hallmark of science is the parallel hand-in-hand progress that theory and experiment make together, each playing a vital role, sometimes with theory leading the way, other times with experiment doing so. Nowhere is this more true than for the areas of research described in this chapter.

While experimental breakthroughs constantly challenge theorists, the reverse is also true, with theorists suggesting new experimental paths and novel ways to reach exciting regimes where new physics can be explored. For example, the possibility of using Feshbach resonances to achieve new regimes of ultracold physics was suggested by theorists. This proposal led to the creation of molecular condensates and opened the way to one of the most exciting recent discoveries in AMO physics: observation of the crossover between Bose condensation and Cooper pairing of fermions. As a result, there is a new link between atomic and condensed matter physics. Likewise, the idea of using ultracold atoms trapped in optical lattices to study the transition between superfluidity and another quantum state called a Mott insulator originated in theoretical studies. The experimental realization of these states opens the way to exciting new potential approaches to quantum information processes and to the realization of quantum simulators to investigate in detail key problems in condensed matter physics. Nonlinear atom optics and the generation of vortex lattices in BECs, along with many other examples of theory leading experiment, illustrate that in AMO science, there is close cooperation between theory and experiment.

4

Extreme Light

What happens when light is pushed to extremes? Focused laser beams are the best way to concentrate and control energy. Lasers in the next decade will have powers exceeding a million billion watts—equivalent to the entire electrical power consumption of Earth concentrated in a single beam of light for a millionth of a billionth of a second. The enormous electric fields present at the focus of one of these beams completely overwhelm the forces that bind electrons in atoms and molecules, leading to exotic states of matter that are usually found only in stars, hydrogen bombs, or particle accelerator collisions. A different kind of extreme is the wavelength of the light. During the next decade, new facilities will use electron beams from accelerators to create x-ray lasers a billion times brighter than our best present sources. These brilliant x-ray lasers will be used to irradiate complex biological molecules with a brief x-ray flash, thereby capturing crucial details about their shape in order to learn what makes them so efficient as they carry out the processes of life. Still a third extreme is the speed of light. New methods have recently been found to reduce the speed of light to be nearly at rest in a material with very little loss in the process (Box 4-1). During the next decade, scientists and engineers will use these new tools to unravel the inner workings of nature and the nanoworld, while also harnessing extreme light for many important technological areas, from more powerful x-ray microscopes to faster drug discovery, and from alternative energy sources to quantum information.

EXTREME X-RAY LASER LIGHT

In the 20th century, short-wavelength light from synchrotrons or lasers in the ultraviolet or x-ray regimes enabled the visualization of the crystalline structure of proteins, the imaging of cells in three dimensions, the study of the electronic structure of superconducting materials, and the creation of highly excited states of atoms, molecules, and clusters. This work included the use of x rays to uncover the double-helix structure of DNA. Indeed, thousands of scientists worldwide

BOX 4-1
Slow Light

Coherent manipulations of vapors of three-level atoms have led to the observation of remarkable phenomena, such as electromagnetically induced transparency, slow light, and nonlinear optics, at very low (approaching single-photon) light levels. These phenomena originate in the subtle interplay and exchange of optical and atomic coherence. For example, in slow light, optical excitations, which propagate at the speed of light in free space, are reversibly transferred into atomic excitations in a vapor. However, these atomic excitations propagate at a fraction of the speed of light. Once in the vapor, these excitations can be monitored and manipulated. They can be subsequently restored (if desired) to optical excitations. Crucially, the transfer process preserves excitation amplitude and phase information. Envisioned applications include signal processing elements such as delay lines, taps, and bandwidth compressors. As these processes can function at the single-photon level, they also enable new approaches for manipulating and storing quantum information.

Recent work has demonstrated the ability to delay light in optical fibers for applications in fiber-optic communication networks. Slow light would be very useful in all-optical routers, which are used in communication systems to direct information from one point to another. Current routers convert optical information into an electronic form (a so-called communication bottleneck), while an all-optical router would eliminate the optical to electrical to optical conversion and greatly speed up the process. An all-optical router would require an optical buffer—a device that would function as temporary optical storage—to synchronize data packets effectively. A slow light device would accomplish this function. Researchers have recently demonstrated a more than 300-fold reduction in the group velocity of an optical pulse propagating on a silicon chip by using an ultracompact photonic integrated circuit with a silicon photonic crystal waveguide. Many view slow light as an important aspect of our future capability to process and transport information optically: Photonic crystals may be the key to that future.

continue to use short-wavelength light for an astonishing array of applications in basic physics and chemistry, in biology, in materials science and engineering, and in medicine.

New advances in atomic and optical physics are creating brilliant bursts of x-ray beams with laserlike properties. These bright, directed x-ray beams can be focused to the size of a virus and are fast and bright enough to capture the complex dance of atoms within molecules or—even faster—the fleeting motion of electrons within atoms and molecules. These extreme strobe lights, with x-ray vision, will provide a direct view of the electronic and structural changes that govern biology and nanoscience at the molecular level. Scientists have never had such a window through which to explore the nanoworld.

Advanced x-ray sources will be developed through the combined efforts of scientists in universities, national laboratories, and industry. Their scale will range from tabletop systems designed for very short pulses of soft x rays to large national facilities capable of generating brilliant high-energy pulses of hard x rays. The progress in both pulse duration and pulse brightness is dramatic, both for x rays and for visible lasers, as shown in Figure 4-1.

TABLETOP SOURCES OF X RAYS

Table-sized sources of x rays, such as the ones used for medical or dental x rays, have been around for a century. Although tremendously useful for medical diagnostics, a major disadvantage is that they produce weak pulses of quite long duration (see Figure 4-1 for source brightness comparisons). This combination—dimness and duration—means that the tabletop x-ray machines of the 20th century are useless for viewing very small samples or fast events such as chemical changes in a molecule, which become blurred if the x-ray pulse is longer than a trillionth of a second.

Two new types of table-sized soft x-ray lasers have been developed in the past 10 years. (Box 4-2 discusses the difference between soft and hard x rays.) The first uses ionized atomic plasmas as the lasing medium to generate highly monochromatic and directed laser beams at wavelengths from 11 to 47 nm. Figure 4-2 shows one such laser. These devices are already in use for high-resolution spectroscopy and microscopy, and their ruggedness and relatively low cost may lead to industrial applications in next-generation microchip mask manufacturing or in nanomachining.

Another exotic new source of coherent, laserlike beams at soft x-ray wavelengths uses an extreme version of nonlinear optics (see Box 4-3). By focusing an intense femtosecond laser into a gas, the electrons in the gas atoms are driven so nonlinearly that high harmonics of the fundamental laser are emitted as coherent, laserlike beams at short wavelengths. The photon energies released by this high-harmonic generation process can span the entire ultraviolet and soft x-ray region of the spectrum, up to photon energies of kiloelectronvolts and higher. During the next decade, these tabletop extreme light sources may be used for compact microscopes capable of imaging, with unprecedented time resolution, the complex nanoworld within a single cell or for probing the behavior of materials and interfaces on nanoscale dimensions. Furthermore, the high-harmonic x-ray emission process itself is extremely brief, much shorter than the laser pulse duration. This means that by using very short laser pulses, 5 femtoseconds or less, it is possible to create x-ray beams with ultrashort, subfemtosecond (or attosecond) durations. Pulses generated in this manner are the shortest strobe lights that we can generate to date—short enough to capture some of the fastest events in atomic and molecular science. As we will read in Chapter 5, studies using these pulses will add to our fundamental knowledge of how electrons, atoms, and molecules respond to light.

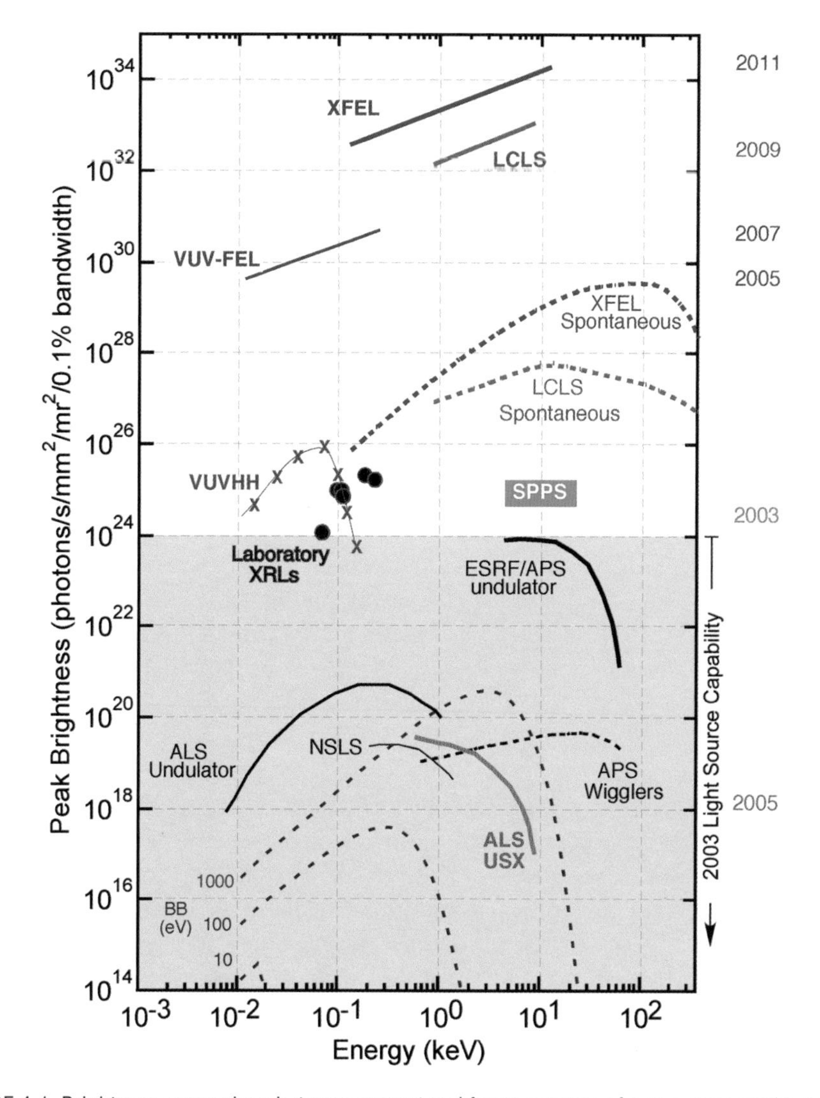

FIGURE 4-1 Brightness comparison between current and future sources of x rays generated in laboratory x-ray lasers or at accelerators. Sources shown are the National Synchrotron Light Source (NSLS) at Brookhaven National Laboratory; Advanced Light Source (ALS) at Lawrence Berkeley National Laboratory; Advanced Photon Source (APS) at Argonne National Laboratory; European Synchrotron Research Facility (ESRF); Subpicosecond Pulse Source (SPPS) at the Stanford Linear Accelerator Center (SLAC); Linac Coherent Light Source (LCLS) at SLAC; the vacuum ultraviolet XFEL at the TESLA Test Facility at DESY, Hamburg, Germany (VUV-FEL); and the future European X-Ray Laser Project XFEL. Years on the right edge of the diagram denote approximate commissioning dates. SOURCE: R.W. Lee, Lawrence Livermore National Laboratory.

BOX 4-2
X Rays—Soft or Hard?

X rays originally referred to radiation that emerged after exciting very tightly bound electrons in atoms. These days, the term "x ray" can mean any source of radiation between ≈250 eV (4 nm) and 50 keV (0.02 nm). Sometimes the definition extends to encompass vacuum ultraviolet radiation at 50 eV (20 nm). X rays are classified as soft or hard depending on the energy of the x-ray light particle (photon); the energy determines how far the photon will penetrate matter. Higher energy photons (harder x rays) travel deeper into matter before they are absorbed. The commonly accepted dividing line between soft and hard is at about 1 kiloelectronvolt of photon energy, or around 1 nm wavelength. To get a feel for what this means in practice: Dental x rays and radiation therapy x rays are hard; solar flares produce both hard and soft x rays; and the gaseous plasmas in arc lamps and welders produce mainly soft x rays. These light sources may be used for next-generation microlithography, where the energy needs to be absorbed in a very precise nanopattern in a very thin layer (see Figure 4-2-1).

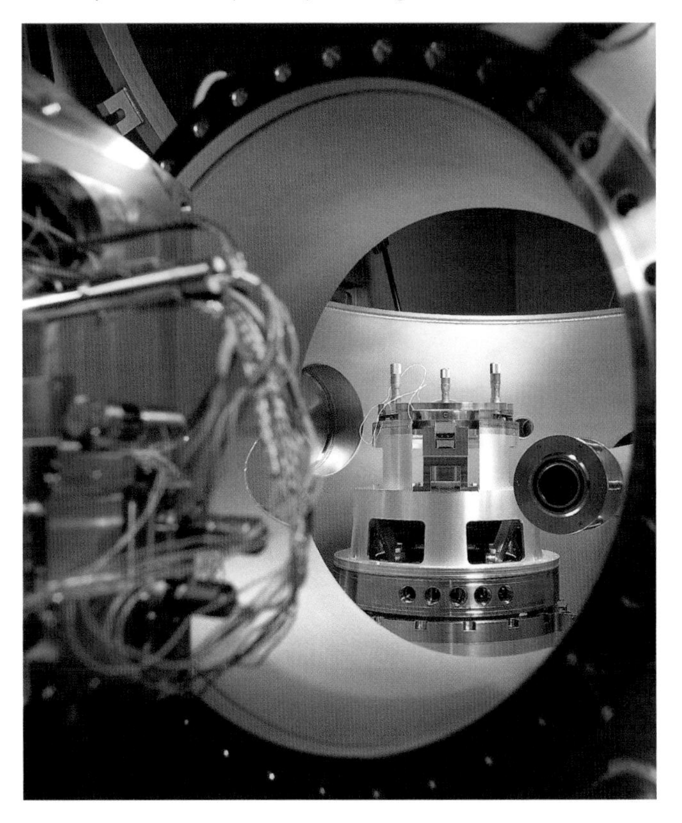

FIGURE 4-2-1 A vacuum chamber for processing microchips using soft x rays. The vacuum prevents air from absorbing the x rays and also keeps the microchips clean. SOURCE: Carl Zeiss A.G.

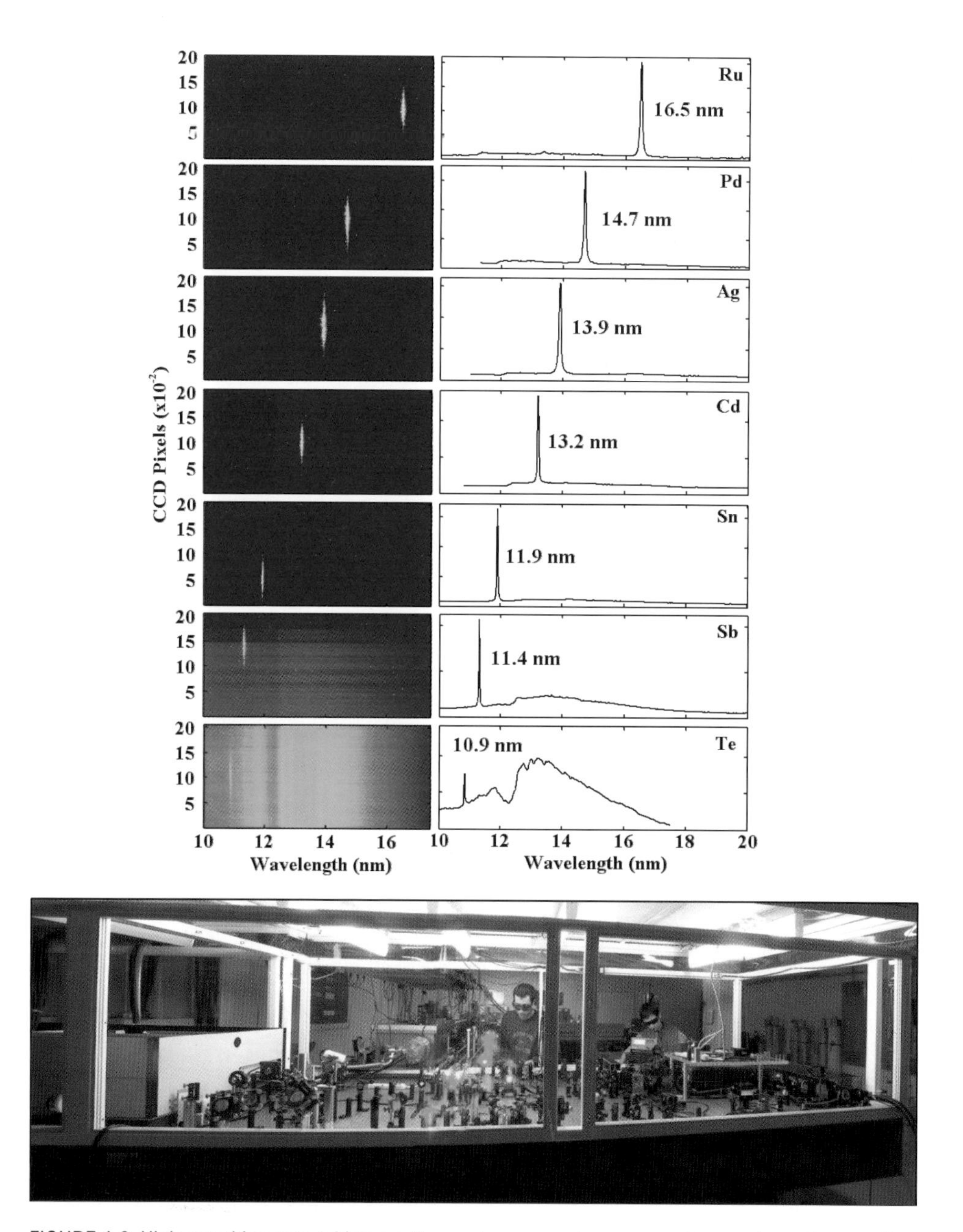

FIGURE 4-2 High-repetition-rate, tabletop soft x-ray lasers may become workhorses in next-generation photolithography, metrology, and other nanotechnology applications. Upper charts show the soft x-ray emissions from the tabletop x-ray laser at Colorado State University, shown in the lower picture. SOURCE: J.J. Rocca, Colorado State University.

BOX 4-3
High-Harmonic Generation

In the 20th century, optical scientists harnessed a remarkable property of certain transparent crystals: When laser beams containing photons of one frequency, such as those corresponding to visible or infrared light, pass through these crystals, they fuse together in twos and threes to form new photons whose frequency is the sum of the photons in the original group. This leads to doubled or tripled frequency of the original laser and a correspondingly shorter wavelength. This traditional nonlinear optics extends laser light throughout the visible, infrared, and ultraviolet regions of the spectrum.

Extreme nonlinear optics happens when we turn up the focused laser power and electrons are literally ripped away from atoms by the laser field. These same electrons then gain energy from the laser field and liberate this accumulated energy as a high-energy photon when they are forced to recollide with the atom. This process, called "high-harmonic generation," is a totally new photon conversion process. Instead of the two or three laser photons that are added together in traditional nonlinear optics, high-harmonic generation combines tens to hundreds of visible laser photons together, to generate laserlike beams with photon energies up to the kiloelectronvolt regime.

Although we can think of high-harmonic generation as an extreme version of nonlinear optics, this process of stripping away electrons from an atom is really a quantum phenomenon and is deeply affected by the quantum wave nature of the electrons as they move under the influence of the laser. Figure 4-3-1 shows the quantum wave of an electron as it is being gradually stripped away from its parent atom and accelerated by a strong laser field.

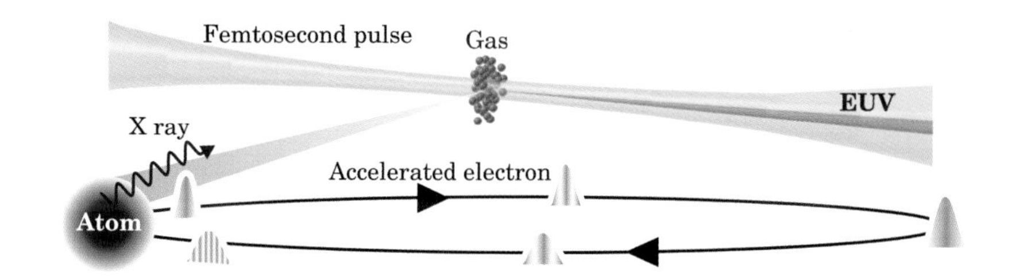

FIGURE 4-3-1 Simple and quantum pictures of high-harmonic generation. *Top:* An electron is stripped from an atom, gains energy, and releases this energy as a soft x-ray photon when it recombines with an ion. *Bottom:* Two-dimensional quantum wave of an electron is gradually stripped from an atom by an intense laser. Fast changes in this quantum wave lead to the generation of high harmonics of the laser. Reprinted with permission from H.C. Kapteyn, M.M. Murnane, and I.P. Christov, 2005, Extreme nonlinear optics: Coherent x rays from lasers, *Physics Today* 58 (3). Copyright 2005, American Institute of Physics.

Many experimental and theoretical challenges will have to be overcome by scientists and engineers during the next decade to reap the full benefit of these tabletop x-ray sources. With further advances in extreme nonlinear optics, these sources have the potential to be much brighter and to span a larger photon energy range while generating designer femtosecond- or attosecond-duration pulses. Hybrid x-ray

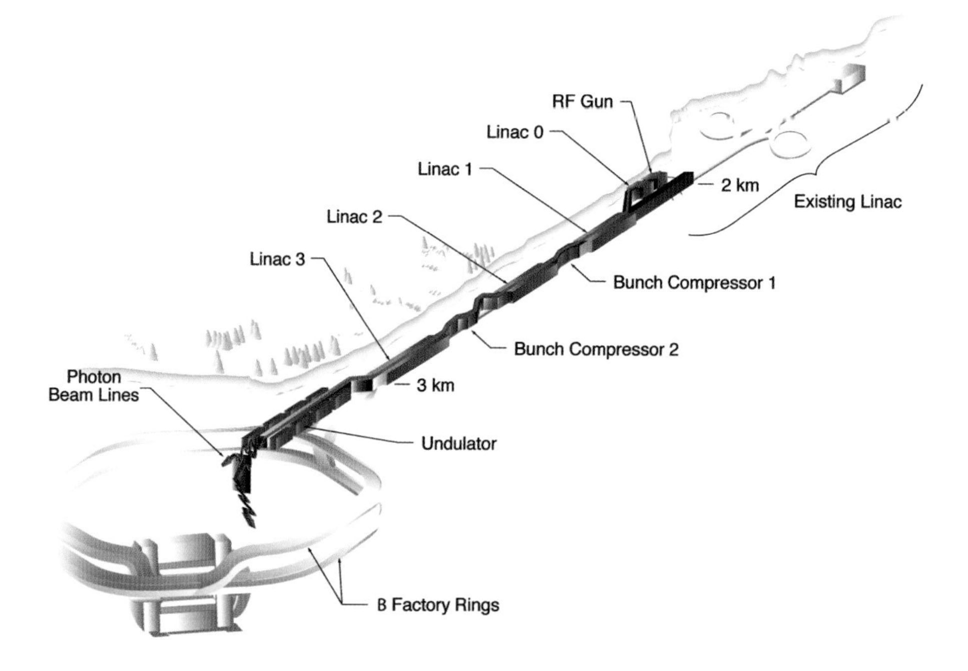

FIGURE 4-3 Artist's depiction of LCLS showing the underground labs and the path of the XFEL beam. SOURCE: Stanford Linear Accelerator Center.

sources that combine the benefits of different kinds compact x-ray sources have just been demonstrated, paving the way for further advances in extreme light.

EXTREME X-RAY LIGHT SOURCES AND THE WORLD'S FIRST X-RAY LASER FACILITY

The largest x-ray laser currently under construction in the United States and indeed the world, and the first scientific user facility for x-ray lasers, is the Linac Coherent Light Source (LCLS) x-ray free-electron laser (XFEL) at the Stanford Linear Accelerator Center (SLAC)[1] (see Figure 4-3). As described in Box 4-4, the LCLS (and its sister projects being planned in Europe and Japan) will produce an x-ray laser beam of sufficient brightness to illuminate and capture pictures from single biomolecules (see Figure 4-4). The molecular architecture of a complex biomol-

[1]For further information, see DOE/BES, *LCLS: The First Experiments* (2000), and NRC, *Frontiers in High Energy Density Physics*, Washington, D.C.: The National Academies Press (2003).

BOX 4-4
Structural Biology: Novel Approaches to the Study of Macromolecular Structure Using X-Ray Free-Electron Lasers

X-ray photons from synchrotron storage rings have revolutionized structural biology. Today one can use synchrotron-based crystallography to study the intricate details (at the atomic level) of very complex biological assemblies that have been formed into a crystalline sample. These developments have primarily utilized the high average brightness and broad spectral range of current-generation synchrotron x-ray sources. These x-ray sources, however, are based on electron storage rings and are therefore constrained in the type of x-ray radiation that can be produced. For example, x rays from today's typical third-generation x-ray sources have relatively long pulse durations, ranging from tens to hundreds of picoseconds. Moreover, there are a relatively small number of coherent photons in the hard x-ray regime. Therefore, use of a crystal containing millions of copies of the molecule (thus greatly amplifying the magnitude of the scattering effect) is required to determine the structure. These limitations result in two important scientific barriers. First, with present-day x-ray sources, structure determinations with atomic or near-atomic resolution can only be performed on biomolecules (for example, proteins) that can be crystallized. Second, only static or very slowly evolving structures can be measured; however, the processes underlying biological function involve dynamically evolving molecular structures.

In contrast, the extreme XFEL sources that use linear electron accelerators have the potential to produce much shorter and more brilliant x-ray pulses. These extreme x-ray laser sources take advantage of true laser amplification to generate coherent beams that are exquisitely directed and focusable. The promise of these so-called fourth-generation XFELs, such as the LCLS at SLAC, is to open up a completely new realm of x-ray science, enabling a new era of single biomolecular and nanostructure determination as well as the ability to study structural dynamics in materials and chemical/biological systems.

The challenge of developing completely new approaches for noncrystalline, atomic-level molecular imaging is, however, formidable. Many problems must be overcome for new approaches to succeed and become practical. Therefore, a multidisciplinary approach is mandatory, requiring a collaboration of the finest minds and most talented experimenters in structural biology, AMO physics, mathematics, statistics, laser science, and accelerator physics to accomplish the integration of theory and simulations, novel sample injection schemes, high-speed x-ray detection, and new algorithms for data analysis and visualization. The benefits of success for science and society are enormous. We could rapidly and routinely study all biomolecules, including those that are difficult or impossible to crystallize. Freed from the limitations imposed by crystallization, scientists will be able to study many membrane-bound proteins as well as the large molecular machines responsible for many aspects of cellular function.

ecule can be determined from a series of such pictures. While this alone would be an extraordinary accomplishment, XFELs hold additional promise. Unlike current synchrotron x-ray sources, XFEL light will also produce very short bursts (tens to hundreds of femtoseconds) of brilliant x-ray light. In fact, the bursts can be shorter than the movement of the atoms making up the biomolecule and eventually short enough to capture the molecular structure before the molecule explodes as a result of the bright x-ray flash. Furthermore, if the molecule is involved in some chemical

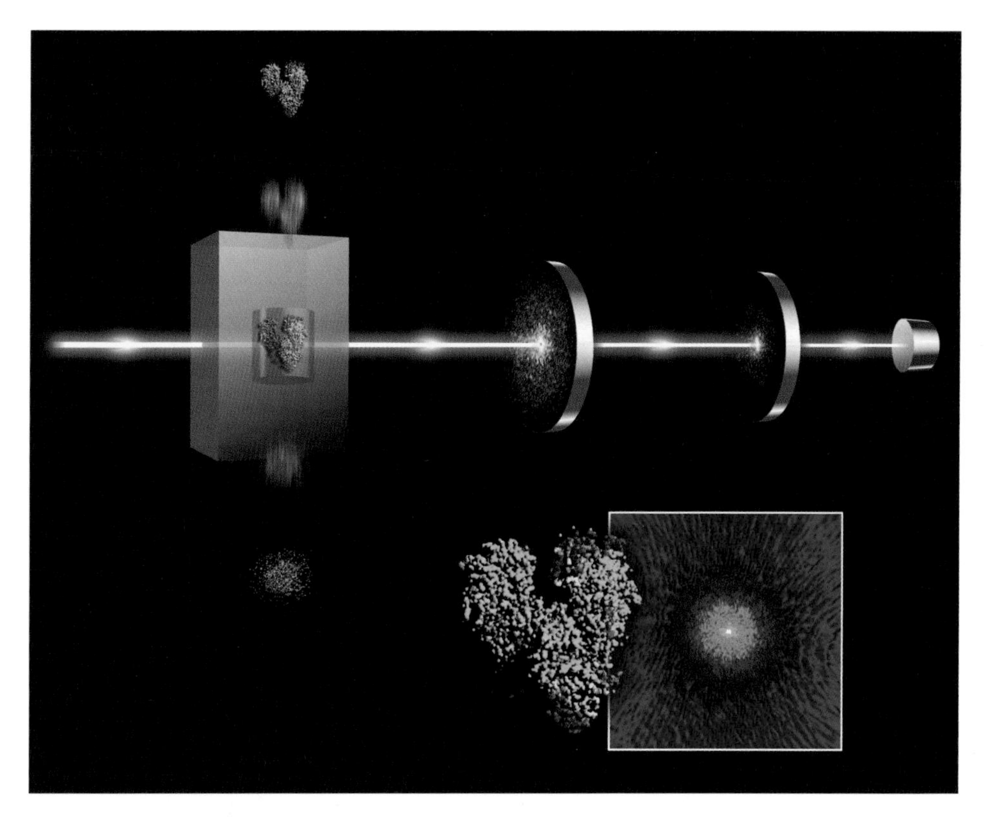

FIGURE 4-4 Single-molecule diffraction by an x-ray laser. Individual biological molecules fall through the x-ray beam, one at a time, and are imaged by x-ray diffraction. An example of the image is shown in the inset. SOURCE: H. Chapman, Lawrence Livermore National Laboratory.

reaction (for example, a chlorophyll molecule as it harvests light), it may be possible to chart the course of all of the atoms in the molecule as it changes.

AMO Contributions to Single-Molecule Imaging

Several fundamental challenges of time-resolved, single-molecule imaging involve AMO physics. The very high radiation damage to a single biomolecule from a focused x-ray laser beam of some trillion x rays is far beyond anything known in protein crystallography. Even the fundamental mechanisms of damage at such high intensities are not well understood and relate to basic questions in AMO physics such as these:

- Are there important new nonlinear damage mechanisms?

- Can the coherence of the x-ray laser change the character of the damage?
- Can we find ways to lessen the effects of radiation damage on imaging by shortening the duration of the laser pulse or by changing the properties of the beam?

These issues affect the quality of the x-ray diffraction pattern that will be analyzed to obtain the molecular structure. It is known that the initial interactions between the x-ray beam and the molecule involve ionization (removal of negatively charged electrons) and that the accumulating positive charge generates a tremendous force within the molecule that gives rise to what is termed a "Coulomb explosion" (see Figure 4-5). What has been realized recently is that if extremely short x-ray laser pulses (tens of femtoseconds or less) with sufficient brilliance per pulse ($\sim 10^{12}$ photons) are used, then an individual x-ray diffraction pattern could be recorded from the macromolecule in the gas phase before radiation damage manifests itself and ultimately destroys the molecule by literally blowing it apart. These results come from theoretical simulation of the complex behavior of the atoms within the molecule before, during, and after absorption and scattering of the x rays and the subsequent Coulomb explosion. There are a number of factors that influence the behavior of molecules under such extreme conditions: Improved theory and simulation techniques need to be developed to better understand them and how they may limit accurate structure determination. The hydrodynamic codes used in understanding plasmas and nuclear events are being adapted to model these "molecular explosions." These simulations need to be benchmarked against experimental data as soon as such data become available.

To determine the three-dimensional structure of a noncrystalline biomolecule using LCLS, a large number of individual two-dimensional x-ray diffraction pat-

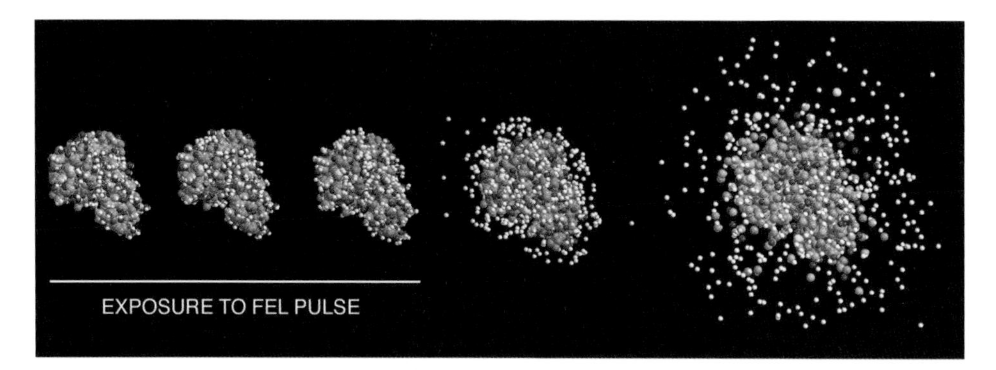

FIGURE 4-5 Artist's depiction of a Coulomb explosion of a protein molecule (lysozyme) exposed to the focused pulse of an XFEL. With a very short x-ray pulse (indicated by the white line), atomic positions remain virtually unchanged during the exposure. SOURCE: J. Hajdu, Uppsala University, Sweden.

terns must be recorded, classified, and averaged. Owing to the relatively weak signal that will come from the scattering from a single biomolecule, all other background contributions must be minimized. One concept being developed is to inject the biomolecules as a molecular beam into the XFEL beam. This injection process is already used and known to work for determining molecular weights using mass spectrometry. In the bioimaging experiment, thousands of two-dimensional images would be recorded sequentially from individual molecules, their orientations determined, aligned, and averaged to produce a single three-dimensional molecular diffraction pattern. Because the individual two-dimensional patterns will be relatively weak, new reconstruction algorithms must be developed that work at the minimum possible signal-to-noise levels. Ultimately, the three-dimensional diffraction pattern will be converted into the three-dimensional molecular structure.

Single-molecule imaging must meet and overcome some formidable challenges. To reduce background, a single molecule must be held in containerless packaging so that only the sample will be imaged. Techniques developed for electrospray can be adapted to the injection of single biomolecules into vacuum. Equally important are molecular dynamics simulations to study how the biomolecules behave under these high-vacuum conditions and how the water structure on their surfaces or other structural elements are affected. There is also significant advantage if, rather than recording images from randomly oriented biomolecules, one can use physical methods to induce alignment (or partial alignment). For example, as discussed in Chapter 5, very strong laser fields have been used to simultaneously force all three axes of a small molecule to align along given axes fixed in space and inhibit the free rotation. The adaptation of these techniques to macromolecules is an area of strong interest.

TESLA Test Facility Early Results

With the recent start of operations of the soft XFEL in Hamburg, Germany, measurements with lower-energy x-rays will soon become possible. The regime of such high-peak-power x-ray pulses has never been accessed before and, while the physics of models and simulations appears to be correct, such experiments are very interesting as they will provide the first direct experience relevant to the eventual use of LCLS for atomic resolution imaging of nonperiodic materials. Figure 4-6 shows an example of single-pulse imaging using vacuum ultraviolet (VUV) light.

Inner Shell Atomic Multiple Ionization

The LCLS x-ray laser beam will be the first x-ray source in history to be able to generate the same extreme focused powers that can be accessed by current-

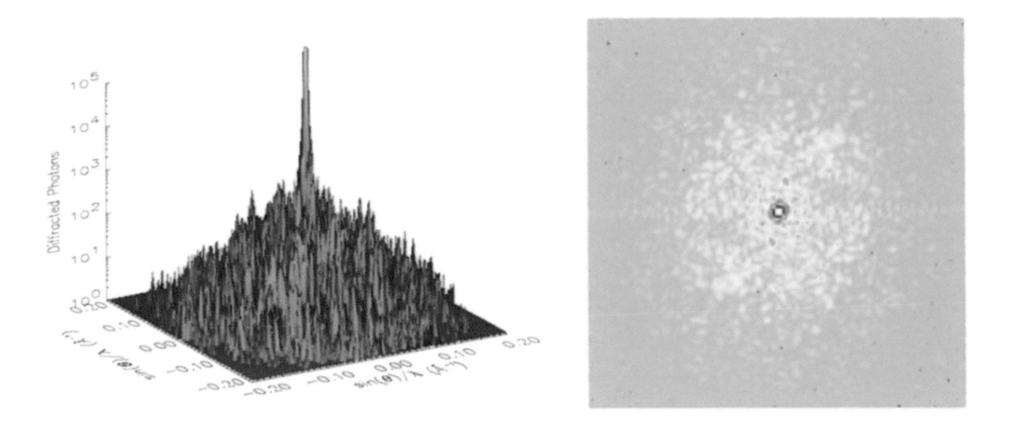

FIGURE 4-6 Simulated section of the three-dimensional pattern from a nonperiodic protein sample (rubisco) processed from 10^6 molecules using LCLS parameters (10^{12} photons; focal spot: 0.1 μm; λ: 15 nm; resolution: 25 nm). SOURCE: Keith Hodgson, Stanford Synchrotron Radiation Laboratory.

generation high-powered lasers. While we can make predictions about what happens to matter as we turn up the x-ray powers to such extreme levels, we also expect to initiate unusual physical phenomena that have not previously been studied, or even observed, anywhere on Earth or in the universe. Chief among these new physical effects is rapid multiple ionization of the most deeply bound electrons. This will lead to the creation of "hollow" atoms and ions—that is, species with two or more electrons missing from the highly bound region at the very center of the atom, next to the atomic nucleus. In the past, such atoms have been produced in high-energy ion-atom collisions, where the ability to study them in detail has been limited because they are created at random times by collisions. Small numbers of hollow atoms have also been produced by an extremely weak process in which a single x-ray photon causes the ejection of the two most tightly bound electrons in the atom, and hollow atoms of lighter elements such as lithium have been produced using light from a combination of lasers and synchrotrons. Such atoms store enormous amounts of potential energy and represent extreme matter in a truly exotic form. The decay of these hollow atoms also provides considerable insight into the correlated motion of the electrons remaining in the atom.

The creation of large numbers of hollow atoms on demand using photons has been impossible to date. Although the intense fields created at the focus of high-powered visible lasers are truly enormous, these lasers tend to preferentially strip away the most weakly bound electrons from the atom—much like peeling an onion. In contrast, the LCLS will produce hollow atoms directly from neutral atoms through inside-out absorption of multiple x-ray photons. This relies on three properties of the x-ray laser, which exist in combination only in this new tool—namely, the short wavelength of the photons, the brevity of the pulse, and the high pulse intensity. The short-wavelength x rays have a much stronger effect on tightly bound "core" electrons than on the weakly bound outer "valence" electrons in an atom. Almost all of the x rays absorbed by the atom remove the inner shell electrons rather than the valence electrons. This property, of course, is shared with all other x-ray sources—inner shell photoemission is a well-known x-ray phenomenon. However, in all experiments to date, the hole left in the atom by the departing electron is filled in so rapidly that any subsequent probing of the atom will not see the hole but will only see the absence of the outer electron that filled the core hole.

This is where the very high intensity of the LCLS comes into play. The intensity is so high that the atom can absorb a second x-ray photon and eject a second electron in a few femtoseconds, even before relaxation has refilled the core vacancy left by the first departing electron. The result of all this is that photoabsorption should be quite different at the LCLS due to rapid multiple ionization. This phenomenon is not only of purely academic interest. The dynamics of many other processes, including the single-molecule imaging described above, will be different due to the formation and presence of these exotic doubly excited states. The study of these phenomena is a high priority at LCLS.

X-Ray Nonlinear Optics

Coupled with the high intensities of these new extreme light sources of x rays is their high coherence. Coherence, briefly stated, is the property that makes a wave move in a regular, predictable fashion. The wake of a boat is an example of a coherent water wave; so is a tsunami. On the other hand, the chop that occurs on a lake on a gusty day has very low coherence, since the wave crests are neither regular nor predictable. The most incoherent light comes from thermal sources like the sun. The most coherent light comes from lasers, and the new extreme light sources—both tabletop and large facilities such as the LCLS—produce the most coherent x rays ever made.

Coherent waves can drive matter much more effectively than incoherent waves—think about a surfer! We have already explored high-harmonic generation, which relies on the coherence of a visible laser beam to produce very high harmonics at soft-x-ray wavelengths. What if we replace the visible laser with a coherent source of x-rays? What kinds of superextreme nonlinear optics would be possible then?

The first nonlinear optical phenomenon to be studied in isolated atoms will probably be multiphoton ionization. This is different from the multiple ionization discussed in the previous section: Instead of ionizing multiple electrons, multiphoton ionization refers to the pooling of more than one photon to ionize a single electron from an atom. Two-photon ionization has much in common with second-harmonic radiation, discussed in a previous section of this chapter. Its observation will be a major milestone in the new field of x-ray laser science.

Many other exotic phenomena may follow the development of intense x-ray lasers. High-flux, short-wavelength radiation will have dramatic effects on materials, which might even lead to new kinds of x-ray lasing mechanics in solids.

Summary of Extreme X-Ray Light Sources

In summary, during the next decade, the availability of coherent x-rays from XFELs, together with the potential for a real breakthrough in x-ray imaging, is setting the stage for a unique and privileged period of discovery in x-ray science that will impact structural biology and other fields where nonperiodic nanostructured materials are central. Atomic-resolution imaging of noncrystalline biological materials appears to be feasible, with many of the key concepts being demonstrated experimentally or with detailed models where AMO physics has a strong role to play. Tabletop extreme x-ray sources will allow us to capture the fleeting motion of electrons in atoms, molecules, and solids for the first time, challenging theory and opening up new possibilities for manipulating matter on unprecedented subnanoscale levels. These tabletop extreme x-ray sources will also bring the source to the application, enabling widespread use of next-generation microscopes and spectroscopies to probe materials with unprecedented spatial and temporal resolution.

ULTRAINTENSE LASERS: USING EXTREME LIGHT SOURCES TO HARNESS EXTREME STATES OF MATTER

Think of a laser beam focused onto a spot on a solid surface smaller than the diameter of a human hair. As we begin to increase the laser pulse energy, we first vaporize the spot to create a crater. At still higher energies, the laser continues to heat the vapor from the crater until atoms and molecules explode into electrons and ions, forming an ultrahot, ionized plasma with a temperature of millions of degrees, similar to a star's interior. The laser pulse energy can then be turned up even higher, so that the laser light pushes the electrons and ions around so violently that they accelerate to relativistic velocities close to the speed of light and hardly interact with each other at all. Finally, lasers can now produce pulses of such incredibly high energies that empty space can be ripped apart, to form new matter and light where none previously existed. Figure 4-7 illustrates the rapid increase in laser intensities achievable over the last 50 years.

In the 20th century, scientists explored these exotic states of matter by generating enormous focused laser powers to try to understand extreme states of matter. The

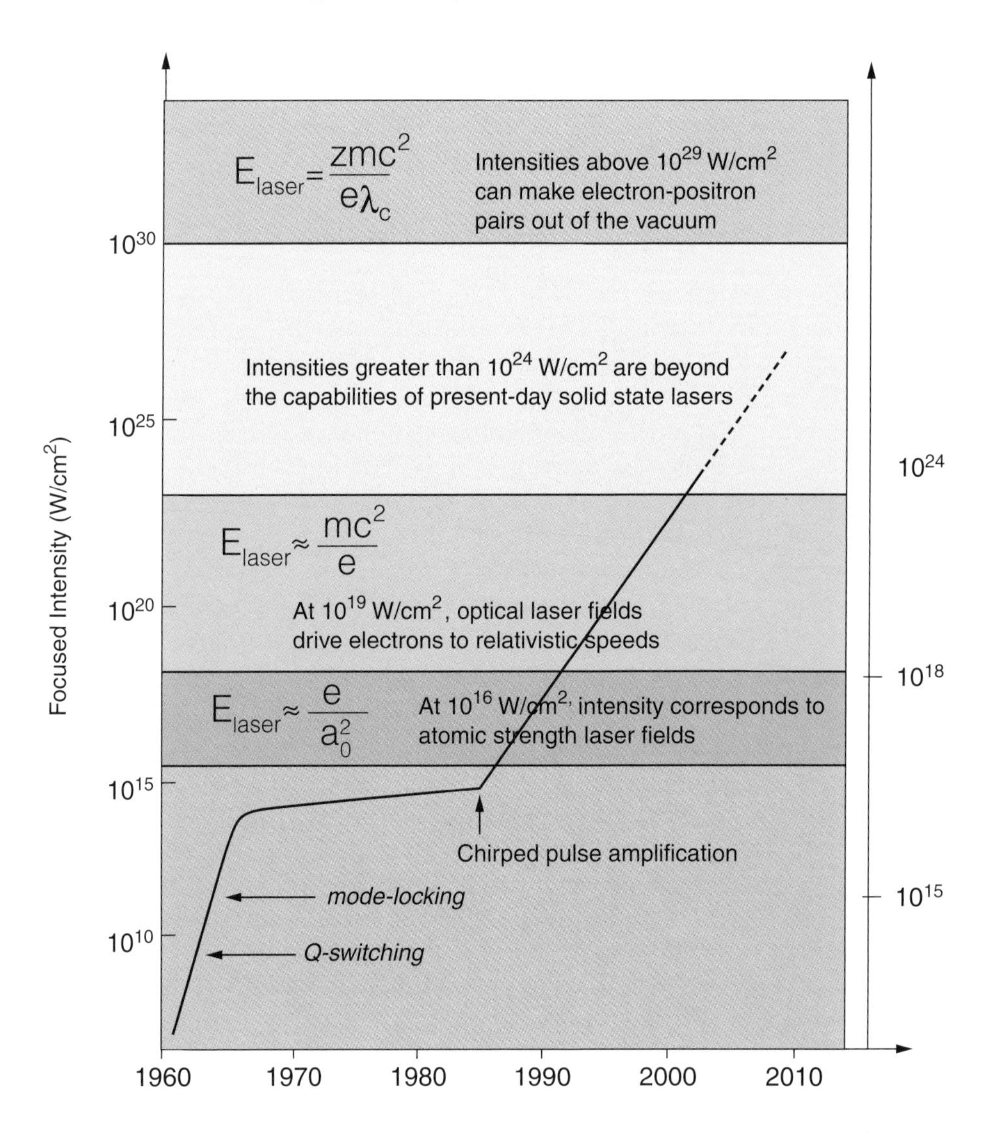

FIGURE 4-7 The exponential increase in achievable laser intensity over the 50-year history of the laser.

challenge in the 21st century is to harness and control such extreme states of light and matter. The biggest accelerator currently being built, the LHC at CERN, requires an accelerator ring 28.5 km in diameter to generate two counter-rotating 7-TV proton beams at great expense and size. Can we design tabletop particle accelerators that can accelerate electrons to gigaelectronvolt energies in a distance no greater than the length of your hand (≈10 cm)? Box 4-5 discusses progress toward this goal. Can we achieve nuclear fusion at the focus of powerful laser beams to harness the atom as a source of clean, abundant energy? Can we use lasers to create energetic x-ray, electron, proton, and neutron beams that will lead to higher-resolution mammograms or allow engineers to predict when an aircraft wing is about to fail?

NIF and Other Large Facilities

At the present time, the National Ignition Facility (NIF) being constructed at Lawrence Livermore National Laboratory is expected to enable scientists to create unique, high-energy plasma conditions in the laboratory that can only be found on Earth during the detonation of a nuclear weapon (see Figure 4-8). The 192-beam, 1.8-million-Joule laser system will address several important scientific questions—some of which are related to producing and understanding basic high-energy-density science, some to the use of fusion as a viable energy source for the world, some to the long-term stability of the nuclear weapons stockpile, and some that will help scientists to understand spectacular astrophysical observations. Lasers with peak power in excess of a thousand trillion watts (1 petawatt) are currently being constructed in the United States, Japan, Britain, France, and Germany.[2] Lasers with powers a hundred times higher than these will be possible in the coming decade. The scientific opportunities enabled by ultraintense lasers to understand, control, and use high-energy-density states of matter are diverse and very exciting.

At the ultrahigh intensities now achievable with the current generation of lasers, enormous electric fields can accelerate electrons to very high energy. Ultrafast, ultra-high-intensity laser production of fast electrons is currently a promising candidate to aid in the ignition of an imploded fusion pellet by externally heating the fusion fuel. This could permit a dramatic leap in the technology of controlled nuclear fusion research. Initial results from Japan and elsewhere are promising. While the prospect of achieving fusion gain high enough for viable energy production is challenging with conventional fusion approaches, fast ignition with intense

---

[2]For further information, see DOE/NNSA/NSF, *The Science and Applications of Ultrafast, Ultraintense Lasers: Opportunities in Science and Technology Using the Brighter Light Known to Man*, Report on the SAUUL Workshop, June 17-19, 2002.

BOX 4-5
Using Lasers to Accelerate Electrons

Directed laser beams provide a very powerful way of concentrating energy efficiently provided this energy can be transferred to a particle such as an electron so that it can be accelerated to teraelectronvolt (1 TeV = 10^{12} eV) energies. An electron in an intense laser beam will behave much like a cork on water as a huge wave passes by, bobbing up and down on the wave but not really going anywhere. More than two decades ago, Tajima and Dawson solved the problem of how to transfer the laser beam energy to an electron: An intense focused laser beam can create a giant fast-moving plasma wave as it passes through a gas and ionizes it. The electron can then surf on the plasma "tsunami" and be accelerated to enormous energies. The laser beam in essence generates a large-amplitude plasma density wave that ripples through the plasma like a wake behind a boat. The longitudinal electric field associated with this density wave can easily be in excess of tens of gigaelectronvolts per meter (1 GeV = 10^9 eV), which is more than three orders of magnitude beyond conventional accelerator technology.

Although experiments that tested this concept produced large gradients for accelerating particles, until 2004 the accelerated electron beams had 100 percent energy spreads with only a small fraction of electrons at high energy. However, in exciting advances independently achieved by three different groups worldwide in the United States, the United Kingdom, and France, the generation of 100-MeV-class electron beams with narrow energy spreads was demonstrated using laser-plasma accelerators. Such high-quality, narrow-energy-spread electron beams are necessary for exploring several scientific frontiers, such as for generating high-brightness x-ray sources, for producing electron and positron beams with energies in excess of 1 TeV, for creating particles from the vacuum, and for testing the fundamentals of quantum and classical electrodynamics.

Figure 4-5-1 shows an experiment in which an electron bunch of around 2×10^9 electrons was accelerated, and the bunch length was inferred to be near 10 femtoseconds. The limitation on the maximum achieved energy to date stems from the fact that the plasma density used in these experiments was relatively high (1-4 $\times 10^{19}$ cm^{-3}), causing the laser pulse to move relatively slowly through the plasma. Much like a surfer on a wave, accelerating particles move forward on the wave and can ultimately even overtake the laser pulse, thus terminating the acceleration and limiting the energy gain.

The next decade will realize gigaelectronvolt- and teraelectronvolt-class electron beams using next-generation lasers, lower plasma densities, and longer guiding distances for the laser.

lasers is more likely to achieve the high gain needed to realize energy production with fusion.

High Energy Density Science: Laboratory for Extreme Conditions in the Matter-Filled Universe

Although plasmas represent the most abundant form of observable matter in the universe, our understanding of this state of matter is remarkably incomplete. This is particularly true of plasmas at very high energy densities. The high-energy lasers described above will provide a laboratory to study the physics of extreme plasma regimes with the promise of learning more about the physics of some of the most important, but inaccessible, plasma environments in the universe—such as

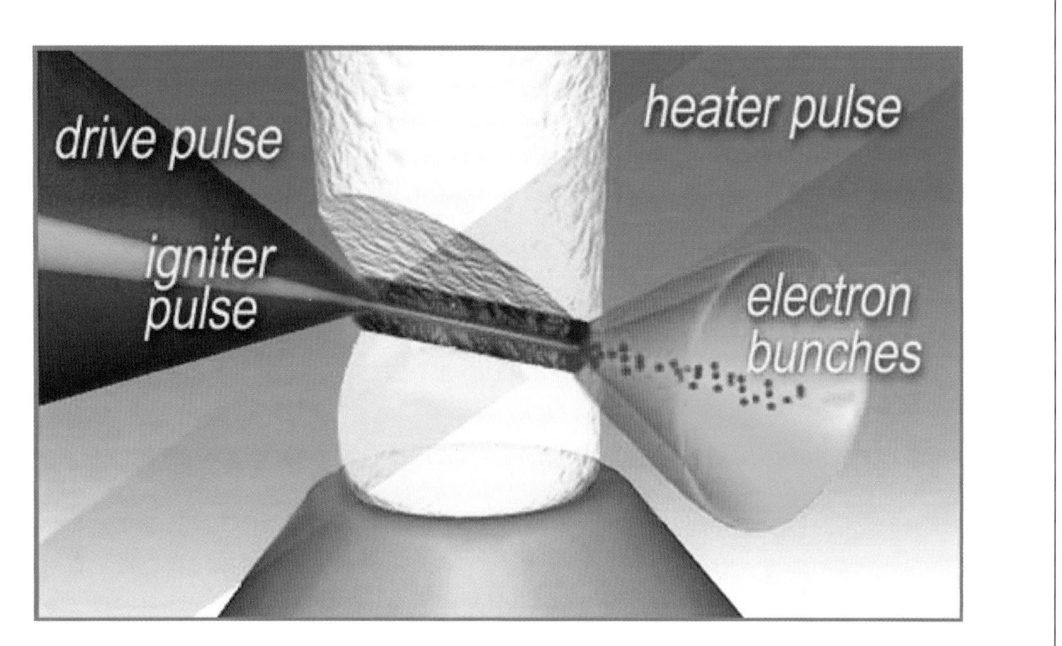

FIGURE 4-5-1 Cartoon of an experiment in which a plasma channel is powered with an 8-9 TW laser, resulting in the production of monochromatic, energetic, low-emittance electron beams with average accelerating gradients near 50 GeV/m. The charge density per megaelectronvolt of the electron bunch observed at 86 MeV is hundreds of times that observed in previous unchanneled experiments, and the electron energy spread is at the percent level. SOURCE: W. Leemans, Lawrence Berkeley National Laboratory.

the cores of neutron stars and white dwarfs or the plasmas near black holes. High energy density (HED) plasmas also exist in nuclear explosions, or in plasmas that might be controlled to produce energy from nuclear fusion. HED science therefore is an important frontier field of modern science. In addition, these HED plasmas may be controlled for application in many important technological problems, ranging from accelerating particles to high velocity to developing new and precise imaging technologies.[3]

Ultraintense lasers provide a controlled means for creating and studying these

[3]For a review of opportunities in HED physics, see NRC, *Frontiers in High Energy Density Physics: The X-Games of Contemporary Science*, Washington, D.C.: The National Academies Press (2003), available at <http://fermat.nap.edu/catalog/10544.html>.

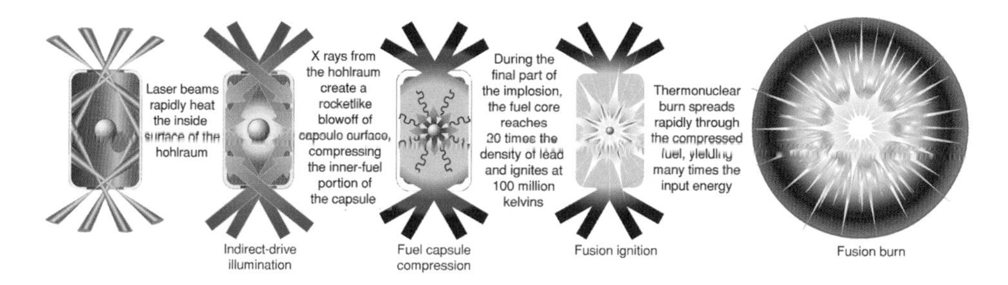

FIGURE 4-8 The mission of the NIF is to produce high energy density conditions and, ultimately, demonstrate fusion ignition through a process called inertial confinement fusion. Laser scientists have successfully executed a laser performance campaign to verify system performance, validating the NIF's ability to amplify, transport, and position beams on target with extraordinary precision and accuracy. These campaigns established world records for laser operation at infrared, green, and ultraviolet wavelengths, setting the stage for the first laser-plasma interaction experiments on the NIF. SOURCE: M. Sherman, University of California, Lawrence Livermore National Laboratory, Department of Energy.

unique states of HED matter. Laser technology has advanced in recent years to the point where light pulses with peak powers of tens to thousands of trillions of watts are possible. These ultraintense lasers can essentially concentrate the equivalent power of the entire electrical grid of the United States onto a spot only a tenth of a human hair in diameter (though only for an instant). It is this concentration of such enormous powers in a laboratory setting that now allows the controlled study of matter in HED states.

One of the greatest challenges in the field of HED science is a theoretical one: to develop conceptual models that can describe the behavior of these new exotic states. Although HED matter exhibits characteristics that range from those expected of a plasma to those that are like condensed matter, it frequently behaves like neither. Intense lasers now allow us to study this behavior and to craft theories that describe it.

For example, a high-energy ultrafast laser can heat solid matter on a timescale much faster than the material expands. This heating at high density produces very high pressure states of matter, in some cases with pressure well above 1 billion atmospheres. Matter in these states is normally found only in the interiors of large planets, dense stars, and nuclear detonations. Thus, controlled laboratory experiments that can inform the study of stars or enhance the nation's security are now possible.

The extremes in temperature that can be accessed with intense lasers now make possible laboratory experiments that could aid in understanding exotic astrophysical events. For example, it is believed that plasmas composed of a mixture of matter and antimatter may exist near black holes. Some day, intense lasers may even permit the creation of small amounts of such hyperenergetic matter in the lab. Other scientific frontiers that will be uncovered in astrophysical research are discussed below.

Accelerating Particles with Light

The preceding sections have just described how the enormous electric fields present in the plasmas created by superintense lasers can accelerate electron beams to multi-GeV energies within a few centimeters.

The Energy Frontier

The high-energy frontier for particle physics will require particle energies well in excess of 1 TeV for studies of fundamental properties of matter. The International Linear Collider (ILC), a superconducting radio-frequency accelerator, has been proposed by the international high energy physics community as a primary tool for such studies and may be built in the next two decades. However, laser accelerator science has advanced very rapidly in the past 5 years, and this holds out the prospect that new accelerator technologies based on lasers may play a part in future high-energy accelerators. There is no clear path for this at present, but the technical problems are understood, and some possible avenues for implementing this vision have emerged.

One possible route to achieve this would be to stage a hundred or more smaller acceleration stages (10-GeV "modules"), each one driven by a synchronized petawatt-class laser. The overall length of such an accelerator would be 200-500 m, a fraction of the distance that would be needed to accelerate a particle to these enormous energies using conventional approaches. The cost would also be a fraction of the projected cost of the ILC. However, there are very significant challenges ahead: To achieve the required high luminosity, high-repetition-rate lasers are required, and to achieve high wall-plug efficiency, a revolutionary new approach in the design and implementation of high-energy laser systems will be required. This is one of the grand challenges for laser science and technology.

The Ultrafast Source Frontier

The bright femtosecond electron bunches that are created using laser-based accelerators can be used as new probes of atoms, molecules, and materials. The electron bunches themselves can be used for time-resolved electron beam diffraction, which is complementary to x-ray diffraction. Such experiments require electron beams with energies of 0.1 to 5 MeV, 10^6 to 10^7 electrons per bunch, low energy spread, and good directionality (emittance). Experiments are under way to demonstrate such beams. A few electrons present in a plasma created by an intense femtosecond laser can receive a boost in energy from an additional laser pulse that enables them to catch the fast-moving plasma wave. By carefully controlling the wave amplitude to avoid wave breaking (and hence self-trapping), simulations indicate that electron beams suitable for electron diffraction experiments could be produced.

Other sources of radiation can also be generated from energetic electron beams accelerated by lasers. For example, broadband coherent THz fields (1 THz = 10^{12} Hz), corresponding to wavelengths of 100 microns, with high fields in excess of 1 MeV/cm could be generated. Such fields can be used for spectroscopy and for probing and/or exciting various materials, such as superconductors, magnetic materials, and nanostructures. Another exciting and important prospect is to demonstrate the next generation of x-ray tubes by crashing the bright electron beam into a target. Because the number of electrons accelerated by an intense laser could be orders of magnitude brighter than currently used in a conventional x-ray tube, the generated x-rays beam could also be very bright and energetic, with polarized photon energies from keV to multi-MeV by proper choice of electron energy, scattering laser wavelength, and geometry. Looking even further into the future, even more intense x-ray beams can be generated using XFELs. However, in some envisioned configurations, these devices require a seed electron beam. If the experimentally demonstrated low energy spread from laser-accelerated electron beams is maintained as the mean energy increases from 100 MeV to a few GeV, and if the electron beam divergence is maintained, intense 100 nm radiation could be produced (10^{13} photons/pulse) from a high-gain XFEL driven by a laser accelerator. Several groups around the world are pursuing this goal. As discussed above, such a compact source of ultraintense, femtosecond x rays would enable many experiments involving molecular, atomic, and biological systems on the natural timescales of atomic motion.

The Intensity Frontier—Sparking the Vacuum

A third frontier is the intensity frontier, where electric field strengths in excess of the Schwinger critical field limit are generated. At this field strength of 3×10^{18} V/m, the vacuum is unstable, and quantum electrodynamics predicts that electron-positron pairs can be spontaneously generated from the vacuum. Indirect experimental verification of this prediction was first accomplished in landmark experiments at SLAC in the 1990s, which used the relativistic shift in the intensity and photon energy of a laser beam colliding with the SLAC relativistic electron beam to reach the required Schwinger limit. The laser intensity required to reach the Schwinger critical field without this trick of relativity is enormous—around 10^{29} W/cm^2. Such high laser intensities are several orders of magnitude beyond the current state of the art. This would require a megajoule-class laser pumping a meter-scale Ti-sapphire crystal amplifier. However, experiments extending the original SLAC results could study this "vacuum boiling" regime in much greater detail. A natural candidate for such an experiment would be a 10-GeV electron beam scattering from a petawatt-class laser beam. This would enable access to the exciting regime of nonlinear quantum electrodynamics by producing effective field strengths that are four to five orders of magnitude greater than those currently achievable directly from lasers. Such petawatt lasers are currently under construction.

High Energy Density Science and XFELs

The XFEL described in the preceding section is expected to have a big impact on the field of HED science. Since x rays can penetrate dense matter, the XFEL can deposit large amounts of energy uniformly over a reasonable volume of material, transforming ordinary matter into warm dense matter similar to the interiors of planets. Other laboratory sources, such as the NIF laser at Livermore, are unable to perform this function well because of the short absorption length for visible or ultraviolet light. The x-ray laser penetration depth also makes it a unique probe of exotic conditions such as those found in the center of superdense plasmas.

The Fastest Pulse: Complementarity Between Extreme Light and Extreme Particle Beam Collisions

A fully stripped uranium nucleus passing rapidly through a high-Z atom at a distance of one hundredth of a typical atomic diameter from the nucleus of the target atom applies to the inner shell electrons of the atom an electric field a million times greater than the field that an electron in the ground state of a hydrogen atom would experience. If this ion is moving at relativistic speeds (see Box 4-6), the field can grow to nearly a billion times the hydrogenic field. Accompanying the electric field is a magnetic field times thousands of times greater than the strongest laboratory DC magnetic field. This electromagnetic pulse completely destroys the normal environment of even the innermost electron. The time duration can be so short that the spectral content of the electromagnetic pulse can extend into the gamma-ray region, sufficient to remove both inner and outer shell electrons as well as to excite the nucleus. The collision can produce a distribution of nearly all excited states of nearly all ionization states of the target. The target responds by

BOX 4-6
New Opportunities in Collisions with Relativistic Heavy Ion Beams and Other New Ion Facilities

The heavy ion facilities at the Gesellschaft für Schwerionenforschung (GSI), Darmstadt, Germany, are entering a period of major facility construction and development that will place in the hands of AMO physicists an unprecedented array of tools for investigating the interactions of relativistic heavy ions with photons and matter (Figure 4-6-1). These facilities will make available to the experimentalist essentially all charge states of all ions up to fully stripped uranium at energies up to 300 GeV per nucleon (relativistic gamma of 300). The fast ions will be used directly for studies of collision dynamics in strong electromagnetic fields and of fundamental interactions between electrons and heavy nuclei up to bare uranium. Alternatively, they will be slowed, after stripping, and stored in two new heavy ion storage rings for spectroscopy and dynamics studies. Finally it will be possible to bring the ions nearly to rest and to cool them in a heavy ion trap for essentially Doppler-free spectroscopic studies of few-electron systems.

A few of the new opportunities include these:

- *Doppler boost.* When relativistic ions circulating in the ring are radiated head-on by lasers in the visible or infrared, they see photons Doppler shifted into the x-ray region and with intensities orders of magnitude higher than the lasers deliver to resting targets. By using a combination of high-resolution x-ray spectrometers and charge-state selected ion detection, precision spectroscopy can be performed on the circulating high-Z ion.
- *High-precision spectroscopy and quantum electrodynamics (QED).* QED contributions to the energies of such systems, such as the Lamb shift, are much greater than the corresponding contributions for low-Z systems. Nearly any charge state of nearly any ion can be produced in the new facility. These ions can be decelerated, cooled, and trapped for precision spectroscopy in the new facility. In situ spectroscopy can also be performed using resonant electron capture (dielectronic recombination) in the electron cooler in the ring, with a precision easily capable of revealing the QED contributions to the energies.
- *Antiatoms.* The facilities will be developed to gradually take over the task of producing low-energy antiprotons for the production of antiatoms. These experiments are presently being carried out with beams at CERN.

emitting a radiation spectrum rich with lines that display physical effects specific to these high energies. It is sufficient to exceed the Schwinger limit described in the preceding section, creating electron-positron pairs.

With the rapid progress of laser and XFEL technology toward ever shorter pulses and harder photons, one must ask where the accelerator collisional approach will meet the laser approach and what information can be expected to issue from each? Collisions produce the shorter time pulses and the harder photons, but for a

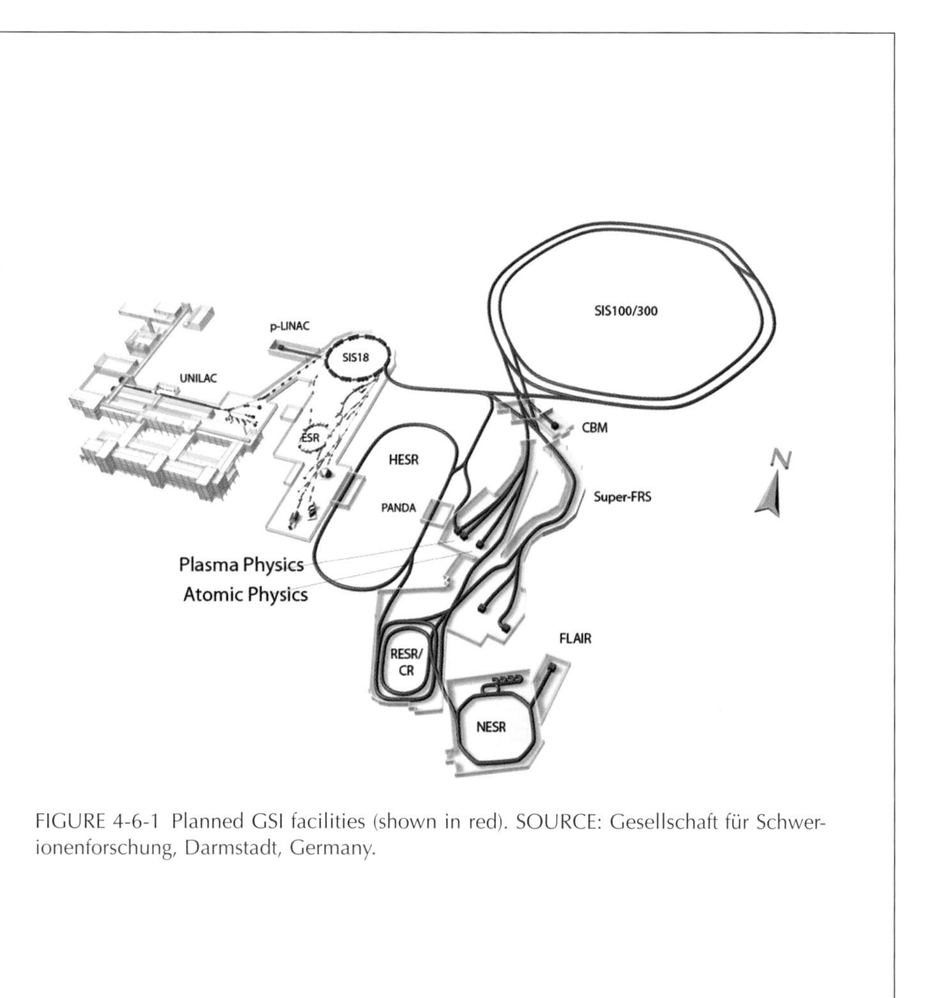

FIGURE 4-6-1 Planned GSI facilities (shown in red). SOURCE: Gesellschaft für Schwerionenforschung, Darmstadt, Germany.

given collision neither of these parameters is under the control of the experimenter. Both the timescale and the frequency spectrum of the photons depend on how close the projectile passes to the target (the impact parameter), and this cannot be dialed in experimentally. In the case of the laser, a macroscopic sample of atoms is exposed to the same pulse length and frequency spectrum, which are under the control of the experimenter. Thus the technologies are complementary, and both are likely to lead to new insights in high-intensity science.

5

Exploring and Controlling the Inner Workings of a Molecule

Can we control the inner workings of a molecule? New 21st-century tools place us on the verge of the new science of coherent control. We will soon not just observe but also control physical phenomena on all of the timescales relevant to atomic and molecular physics, chemistry, biology, and materials science. This new era of control is enabled by key advances in laser technology, which let us generate light pulses whose shape, intensity, and color can be programmed with unprecedented flexibility. Our ability to control the positions, velocities, and relative spatial orientations of individual atoms and molecules has led to a stunning array of precision measurement technologies and devices based on AMO science, leading to an enormous range of experiments that reveal qualitatively new phenomena. In this section, the committee focuses on the emerging ability to observe the inner workings of atoms and molecules on their natural timescales, and to manipulate them to achieve desired effects. Such new capabilities will allow us to visualize the complex motion of electrons and atomic nuclei through the course of chemical reactions, providing new insight into the mechanisms that determine the reaction rates and products. Accompanying our ability to observe is the ability to control: Lasers can now be used to control the outcome of selected chemical reactions. This capability may ultimately develop into powerful tools for creating new molecules and materials tailored for applications in health care, nanoscience and technology, environmental science, energy, and national security.

WHICH TIMESCALES ARE IMPORTANT?

Key events in our lives—graduations, births, or anniversaries—make us acutely aware of the passage of time. Most of our personal time markers are measured in years. At the same time, cell phones, personal digital assistants, and e-mail place ever-increasing demands on our time—seeming to accelerate the pace of our already fast-paced world, where we struggle to preserve even a free nanosecond. In nature, the important time markers span an even broader range—from the dizzying attosecond (10^{-18} s) timescale, the time it takes an electron to orbit an atom, to the timescale of ≈14 billion years, the age of our universe. Fascination with the passage of time is a fundamental aspect of human endeavor, driving us to attempt to understand our world and indeed our universe.

The relevant timescales for the atomic and molecular processes of interest (see Figure 5-1) cover 19 orders of magnitude—a factor of 10 billion billion. The complex folding of a protein molecule can take milliseconds (ms) (10^{-3} s) or longer. On the other hand, a millisecond is a very long time for an atomic collision, unless the atoms are cooled to billionths of a degree above absolute zero, as discussed in Chapter 2. The small molecules in the air we breathe undergo collisions with each other approximately every 100 picoseconds (ps) (10^{-10} s), and they tumble or rotate in space approximately every 1 to 10 ps (10^{-11} to 10^{-12} s). The atoms within

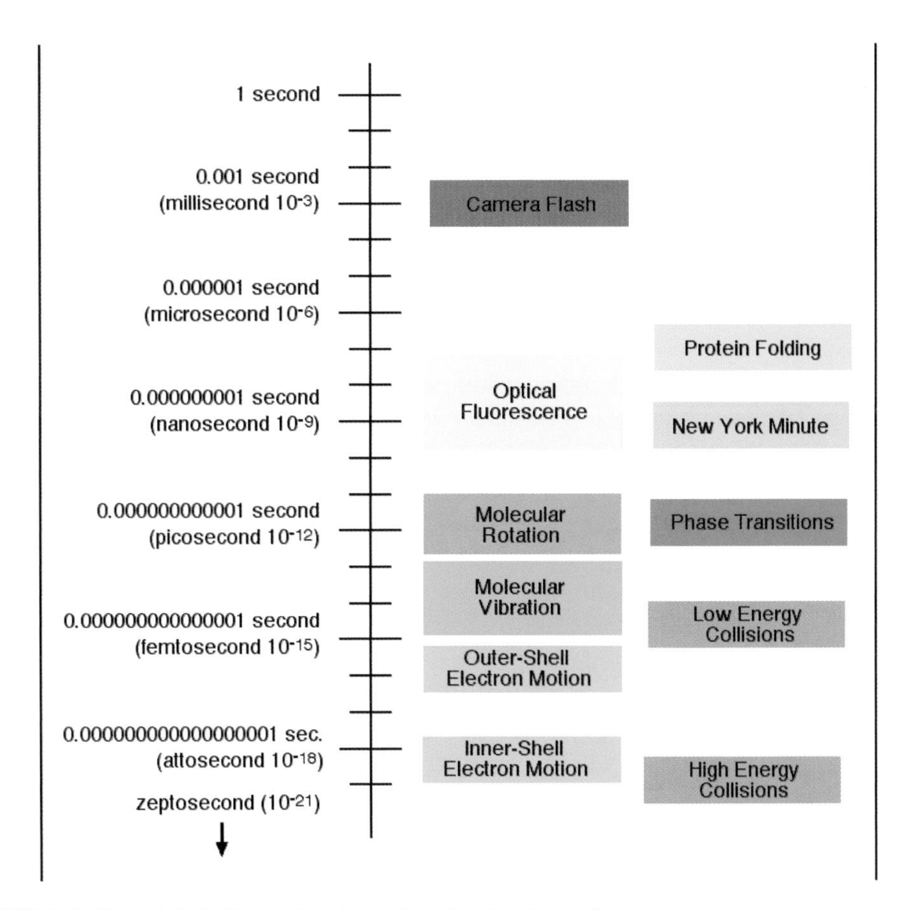

FIGURE 5-1 Characteristic timescales of atomic and molecular motions.

molecules act as if they were balls bound to each other with springs and vibrate with a period of 10 to 1,000 femtoseconds (fs) (10^{-12} to 10^{-14} s) depending on the strength of the spring (i.e., the nature of the bond between the atoms). This period also corresponds to the duration of a typical room-temperature collision between two molecules. Still faster are the orbital motions of the slowest-moving electrons of atoms and molecules, which occur on a timescale of approximately 10 attoseconds (as) (10^{-17} s), while the orbital motions of the fastest electrons of heavy atoms occur in hundreds of zeptoseconds (zs) (10^{-19} s). Much of what we know about atoms and molecules comes from observing motion, starting with the pioneering direct observation of the Brownian motion of dust colliding with air molecules, explained by Einstein in a famous paper published just a century ago. The frontiers in this field as we approach 2010 are to capture motion not just between molecules, but within them, to observe the basic processes of chemistry and biology on the scale of a single molecule.

MOLECULAR MOVIES

Motion pictures help us dissect and understand fast phenomena. About the time physicists were puzzling over Brownian motion in the late 19th century, Eadweard Muybridge used a sequence of photographic exposures of 1 millisecond to prove that a galloping horse sometimes had all four hooves off the ground (Box 5-1). The basic technique established by Muybridge is still used to capture and slow down motion today and helps to illustrate the challenge of capturing the motion within a molecule. A key ingredient in making any motion picture is the ability to freeze the action by recording images with a shutter speed much faster than the motion of the object of interest. In atomic and molecular motion, as elsewhere in high-speed photography, the mechanical shutter has been replaced by a short pulse of light, which acts as a stroboscope. The picosecond or faster processes within molecules require very short pulses which can only be produced by a laser. The speed of the motion we can freeze is limited by the duration of the laser pulse. Thus, the rotational motion of the molecules in a gas cell can be captured by illuminating the gas with laser pulses that are a fraction of a picosecond in duration, while freezing the vibrational motion requires pulses of a few femtoseconds and freezing the motion of electrons as they move about the molecule requires subfemtosecond, or attosecond, laser pulses. Such motion can be captured with laser technology developed during the last decade.

The earliest direct observations of molecular vibrations were performed using a sequence of two pulses, in which the first pulse excited the molecule and the second pulse was used to probe the resulting molecular response as a function of the time interval between the pulses. This pump-probe approach has made it possible to

BOX 5-1
Stopping Time

In the experiments that produced the data at the right of Figure 5-1-1, an x-ray pulse of a few attoseconds in duration is used to knock an electron out of a tightly bound inner orbital of krypton atoms to produce krypton ions. The "hole" in the inner orbital is not stable and is quickly filled by an electron from an outer orbital, with the concurrent ejection of a second electron. How fast does this hole-filling/electron ejection, or Auger process, take place? To find the answer, a second laser pulse of femtosecond duration is introduced at a variable time delay after the first pulse. This pulse can modify the energy of the Auger electron, but only if the electron is still close to the ion, i.e., in the process of escaping. The figure shows a series of electron spectra recorded as a function of the delay of the second laser. If the probe pulse comes before the first pulse, or very long after it (here, "very long" is only 40 fs), the spectra show no signs of the second pulse: The Auger electron has either not yet been produced (before) or is long gone (after). However, at very short times after the first pulse (~0 to 15 fs), a new feature (highlighted in red) is observed in the spectra. This corresponds to electrons with energies modified by the probe laser and marks the appearance of the Auger electron. The Auger process is found to occur with a half-life of 7.9 fs, perhaps the fastest process ever directly measured.

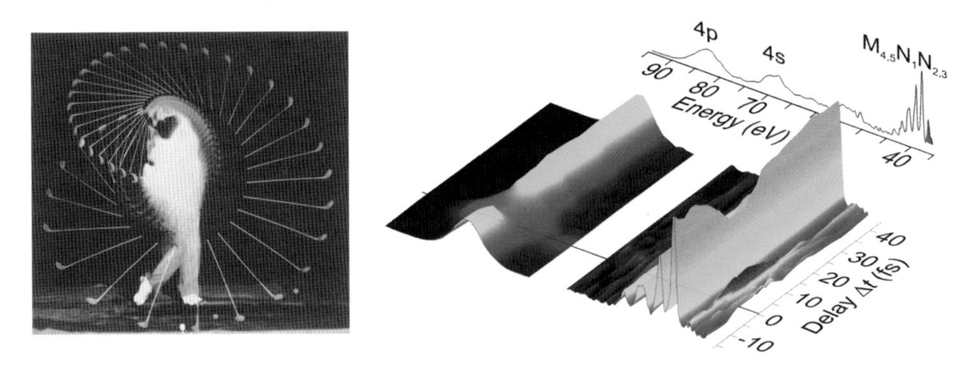

FIGURE 5-1-1 *Left:* In the mid-20th century, Harold Edgerton used a strobed flash with a duration of ~0.000001 s to film a golf swing by Bobby Jones, allowing countless duffers to reassess their games. SOURCE: Palm Press, © Harold & Ester Edgerton Foundation. *Right:* At the dawn of the 21st century, attosecond scientists recorded the evolution of the electronic structure of a krypton ion with a time resolution of ~0.000000000000001 s following excitation by a sub-fs x-ray pulse. Such studies may one day unravel the complex correlated electron motion that drives many chemical reactions and determines the properties of novel new materials. SOURCE: F. Krausz, Ludwig Maximilian University, Munich, Germany, and Max Planck Institute of Quantum Optics.

observe the change in a molecule in solution or in a gas as its atoms stretch apart and contract or break apart to produce fragments. These developments marked the beginnings of the new field of ultrafast chemical dynamics, also known as femtochemistry, and were recognized by the 1999 Nobel prize in chemistry in 1993 (see Figure 5-2). Since this early work, the time resolution of such observations has

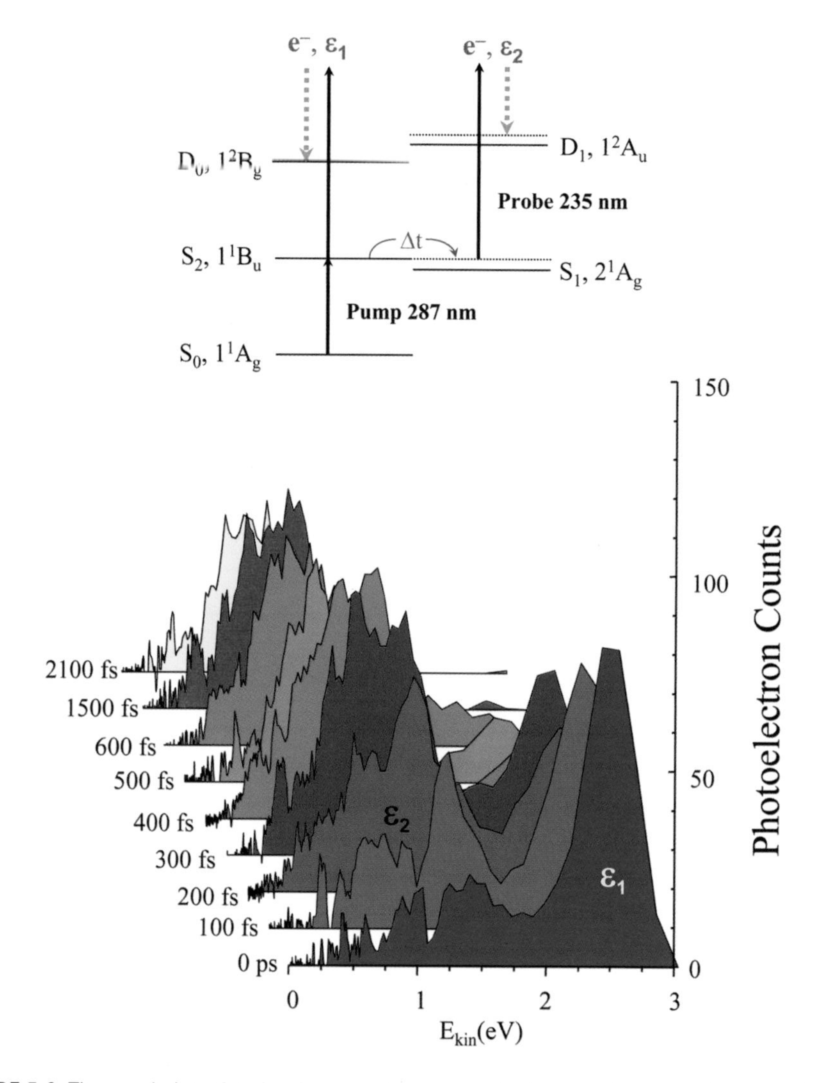

FIGURE 5-2 Time evolution of molecular excited states. Excited states of polyatomic molecules often have a mixed character that results in interesting effects when the molecule absorbs ultraviolet light. Some of these processes can be extremely fast (subattosecond), but others are comparatively slow. This experiment shows the evolution with time of a process that occurs on the picosecond timescale in the excited states of the 24-atom molecule all-*trans*-2,4,6,8-decatetraene. A femtosecond pulse excites the S_2 state, which is mixed with excited vibrational levels of the S_1 state. After a variable time delay, a second femtosecond pulse ejects an electron from the excited state, and the kinetic energy of the electron is measured. This energy spectrum provides a fingerprint of the electronic and vibrational character of the intermediate level. As can be seen above, the spectrum recorded for a time delay of 0.6 ps is very different from the spectrum recorded with no delay. This series of spectra indicates that although the electron is initially in the S_2 state, it rapidly switches over into the S_1 state. SOURCE: A. Stolow, Steacie Institute for Molecular Sciences, Ottawa, Canada.

rapidly advanced from hundreds of femtoseconds to less than 5 femtoseconds, and the increase in the speed of the strobe continues today. At the time of this writing, strobes as short as 100 attoseconds had been reported. On the attosecond timescale, not only the motion of the nuclei of molecules but also the motion of the electrons themselves can be detected. Motion pictures of the electrons will be especially valuable, because electrons form the glue that holds molecules together. Their motion is the fundamental physical basis for chemistry. Why do some atoms bind and others do not? Why do reactions take the time they do, and why do molecules bend one way but not another? Watching the steps in the dance of electrons will provide an enormous wealth of new insight into the mechanisms of chemistry.

THEORETICAL COMPUTATION OF ULTRAFAST MOLECULAR PHYSICS

In this first decade of the 21st century, why is there still a need to do experiments to understand molecular physics? Aren't the forces of nature responsible for molecular binding and atomic motion well understood? Why can't the behavior of molecules simply be calculated based on a first-principles understanding of the underlying physics? If engineers can design and build a commercial aircraft using only computers, and without any test flights, then why can't scientists dispense with experiments? The fact is that the calculation of the chemical dynamics in any but the simplest systems remains an extremely difficult problem. The difficulty is in quantum mechanics itself, which is a far more challenging theory than classical physics. Progress will require both increases in computer power and new theoretical approaches to reduce the size and difficulty of the calculations. Chapter 7 explores the prospects for quantum computers, which may someday permit the simulation of far more complex quantum problems. For now and into the next decade, ultrafast laser experiments and x-ray lasers will lead the way to forge a partnership with theory. One of the most enticing new ways to advance theory through experiments is the emerging field of quantum control.

QUANTUM CONTROL

It is one thing to observe the inner workings of a molecule as it expresses its native behavior; it is quite another thing to exploit what we have learned from these movies to direct or control specific molecular behavior of our own choosing. Such control has been a dream of scientists and engineers since atoms and molecules were identified as the building blocks of matter. It is important not only because it allows us to enhance desired reaction products, minimize by-products, and create new kinds of molecules, but also because of what it can reveal about the fundamental character and behavior of atoms and molecules.

Controlling Chemical Reactions: A Short History

The invention of the laser in the early 1960s raised great hope that this was the tool needed to control molecules. This hope was based on the knowledge that visible and near-ultraviolet laser light can be used to excite the outermost electrons in the molecule into different configurations, which could lead to control of chemical reactivity because these so-called valence electrons determine many of the properties of the molecule. For example, if a vibration could be set up in a specific bond, it was reasoned that bond might be broken selectively, thus achieving mode-selective chemistry. This was truly a revolutionary idea. Synthetic chemistry is more or less like cooking, relying for the most part on heat and on the relative amounts of the ingredients to determine the outcome of a reaction. If a laser could act as a sculpting tool for chemical change, then new chemistry would be possible. Unfortunately, it was soon realized that the coupling among the molecular vibrations was sufficiently strong that before enough energy could be deposited to break a specific bond, much of it had already flowed into different parts of the molecule. Thus, the original attempts at selective bond breaking using lasers were no more effective than simple heating of the molecule, and the laser "scalpel" was behaving more like a laser "blowtorch." Nevertheless, efforts were rewarded by our new understanding of how energy flowed among the internal degrees of freedom of isolated molecules and led to significant advances in nonlinear dynamics and chaotic systems in physics and chemistry.

Quantum Interference: A Route to Quantum Control

Following these initial attempts, new ideas soon appeared that introduced the use of multiple pathways to enhance the desired process and minimize undesired ones (Box 5-2). According to quantum theory, if there are multiple paths between the starting point and the target for a process, then one cannot tell even in principle which path was taken; the paths are said to be "indistinguishable." In such cases, the quantum paths can interfere, much like waves interfering on a beach. Constructive interference means that the quantum waves add, and this leads to an enhancement of the process; destructive interference leads to a diminishment of the process. In principle, if there is sufficient control over the light pulse, it is possible to excite molecules via two or more interfering pathways for which the probability amplitudes interfere constructively for the desired process and destructively for the undesired process. New theoretical methods were developed to discover the optimal combination of pathways for controlling a given process, and from these ideas emerged one of the most effective current methods to achieve this goal: pulse shaping.

BOX 5-2
Quantum Interference

Interference phenomena are familiar in everyday life. For example, when approaching each other, the waves produced by two pebbles thrown in a quiet pond produce an intricate pattern of crests and troughs, with the remarkable property that the height of the wave is higher than that of the individual waves at the crests, and lower at the troughs. The crests and troughs are the result of constructive and destructive interferences, respectively, between the waves produces by the two pebbles. In contrast to pebbles in a pond, which are described extremely well by the laws of classical physics, atoms and molecules are quantum objects governed by the laws of quantum mechanics. Their dynamics is then expressed in terms of "waves of probability amplitudes," or wavefunctions. The knowledge of these waves is extremely powerful in that it permits the computation of everything of interest about the system under study—for example, the probability of finding a particle at a certain location at a certain time. While these waves are rather abstract objects that are fundamentally different from the waves produced by pebbles, the key point here is that their mathematical properties are nonetheless quite similar; in particular, they too can conspire to produce constructive or destructive interferences. For instance, if an atom or an electron propagates past a screen in which two tiny holes have been drilled, these two holes act much like the pebbles in the pond and result in two pathways with waves of probability that will combine downstream, constructively in some regions and negatively in others. According to the laws of quantum mechanics, there is then zero probability of finding the particle where the waves interfere destructively, and an enhanced probability of finding them where they interfere constructively. Similarly, quantum mechanics predicts that if there is sufficient control over a light pulse, it is possible to excite molecules via two or more interfering pathways for which the probability amplitudes interfere constructively for the desired process and destructively for the undesired process.

How Do We Shape an Ultrafast Laser Pulse?

Modern quantum control experiments use a single programmable femtosecond pulse to drive the slowest electrons in the molecule to higher energy orbits. Because these electrons control the bonding in the molecule, control of the electron motion is sufficient to direct the motion of the nuclei. Laser pulse shaping is the key tool needed to carry this out. The pulse is shaped using a physical trick based on the time-bandwidth uncertainty relation: The shorter the pulse, the more colors are contained in it. The pulse is therefore first broken down into its different colors, each of which is then modified by passage through a filter and then recombined to form a complex waveform that excites the orbits of interest. A 10-fs pulse, fast though it is, cannot keep up with the electron motion in real time; it can, however, change the electron orbits and thereby alter the forces that govern the motion of the nuclei, determining the outcome of the chemical process.

How do we choose the appropriate pulse shape to achieve a particular chemical reaction? A sufficiently accurate a priori determination of this pulse shape from knowledge of the quantum structure—that is, the orbits—of the molecule is not possible with current theoretical and computational capabilities for most systems. An elegant solution to this problem was provided by turning the problem around and letting the molecule choose the optimum pulse shape for itself. Sophisticated learning algorithms, such as those inspired by biological evolution, are used to optimize the waveform: The sample is exposed to an evolving series of pulse shapes, while the resulting signal of interest is monitored to determine which waveform maximizes the desired outcome (see Figure 5-3). Because the experiments are per-

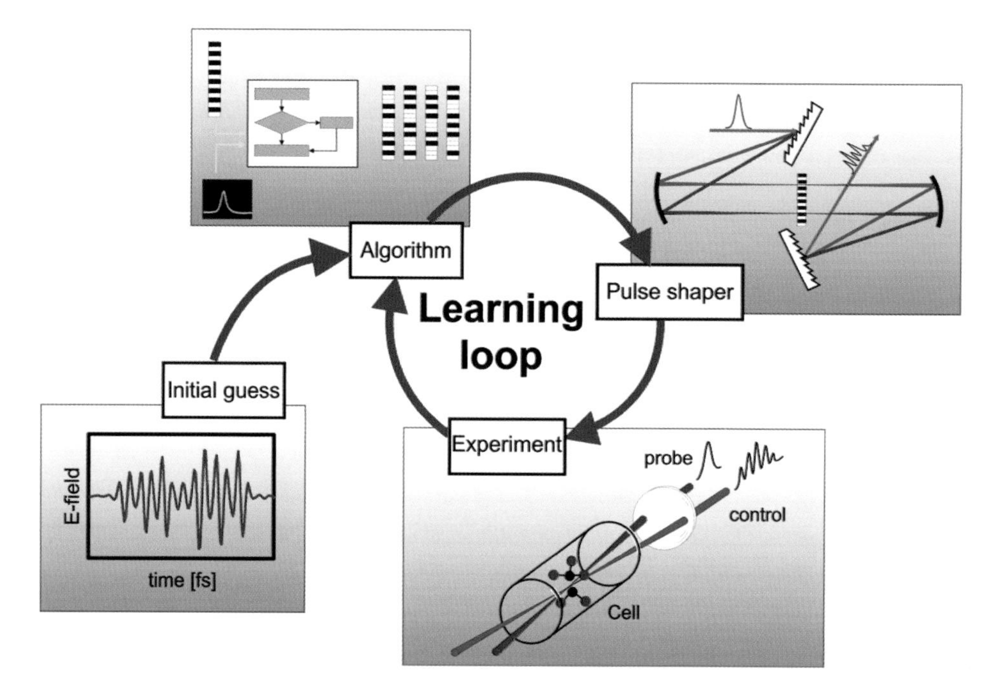

FIGURE 5-3 Letting a molecule solve the problem. A feedback loop can be used to let the experiment choose the optimum shape of the femtosecond laser pulse. In an experiment to optimize the production of a particular product state of a quantum system (e.g., selective fragmentation following the laser-induced breakup of a molecule), the signal from the state is monitored after every laser shot. An algorithm is used to generate a set of initial pulse shapes for the experiment. The femtosecond laser pulse is sent through a pulse shaper to generate the selected waveforms, and these are applied to the quantum system sample. Comparison of the product signal intensities for the different waveforms leads to the selection of a subset of "best" pulses. This subset is then sent through the algorithm again to generate a new, improved set of pulses. The experiment continues around the loop until no further gains can be achieved in the product yield. SOURCE: H. Rabitz, Princeton University, and K. Kompa, Max-Planck-Institut für Quantenoptik.

formed thousands of times per second and the pulse shapes can be modified almost every shot, the molecule can rapidly be exposed to thousands or even millions of different programmed waveforms, and the optimization typically converges on a best pulse shape in seconds to minutes. This learning feedback approach has been used successfully for a number of complex chemical processes, such as maximizing the production of a given quantum state of a molecule, controlling the branching between ionization and dissociation following excitation of a molecule, and controlling the branching between different photofragmentation products. There have even been attempts to control the pathways of energy flow in bacterial light-harvesting proteins.

Quantitative electronic structure and dynamics calculations to predict the optimum laser pulse to control a given process are currently beyond the capabilities of AMO theorists for all but the simplest model systems. Indeed, in general, it is not even a straightforward task for theory to explain why (or if) an experimentally determined pulse is optimal. Such capabilities are ultimately essential to understand quantum control on a fundamental level. In the past decade, tremendous advances in electronic structure theory and computational capabilities have changed the face of molecular physics and chemistry, as recognized by the Nobel prize in chemistry in 1998. Calculations of ground-state equilibrium structures can be exceedingly accurate even for nanoscale systems. In general, however, chemically accurate potential energy surfaces, which describe the energetics of two or more atoms or molecules as they come together in a chemical reaction and rearrange into various products, can only be calculated for systems with a few heavier atoms. Accurate surfaces for excited states of molecules, which are often accessed in quantum control experiments, present even greater challenges. To make matters worse, chemical reactions often involve multiple potential energy surfaces, creating a rapid escalation in the complexity of the problem. Finally, once such surfaces are available, the calculation of the molecular behavior upon them is no simple task.

Nevertheless, theoretical and computational advances in the next decade are expected to lead to major progress on all of these fronts. These advances will not only provide a tremendous boost to the field of quantum control but are also expected to have a broad impact on molecular physics and chemistry and the physical and biological sciences in general.

Aligning Molecules

Selecting the angle of collision between two molecules is yet another way to control chemical reactions. The outcome of a collision depends on the relative orientation and rotation of each collider at the moment of impact. On the molecular scale, the efficiency of chemical reactions and processes like energy transfer depend on how the molecules are aligned when they come together. Similarly, the interaction of light with molecules depends on the relative orientation of the molecules and the direction of the light's oscillating electric field. Control of molecular orientation during collisions or interaction with light could provide considerable control over chemical reactions. Controlling the orientation of free molecules is considerably more difficult than controlling the polarization of light. The latter can be done in many ways, including by using Polaroid sunglasses. As molecules in a gas or liquid spin, rotate, and tumble in space, their orientation is constantly changing. Some "polar" molecules, in which one end is slightly positively charged and the other is slightly negatively charged, will tend to line up in an extremely strong DC electric field, but the aligning force is quite weak, and unless the molecules are quite cold—that is, rotating very slowly—to begin with, the applied field will not be sufficient to keep the molecules oriented (see Figure 5-4). New methods for producing polarized ultracold molecules are discussed in Chapter 3.

Recently, a totally different approach to molecular alignment has been developed in which a very short laser pulse is used to kick the molecules. This kick causes each molecule to rotate faster or slower depending on its orientation and its motion at the time of the kick, much like a weathervane spinning after a brief gust of wind. The net effect is that all of the molecules align with each other a short time after the kick, much like setting a clock to 12:00 a.m. aligns all the hands. The molecules rapidly get out of synch as they continue to rotate, just as the clock hands move apart. In the absence of collisions, however, the molecules come back into phase after a known time delay and are again aligned, just as the clock hands become aligned again at 12:00 p.m. Such recurrences of the alignment will continue to occur until collisions or other external effects erase the effects of the initial kick. One advantage of this approach is that there is no laser pulse or external electric field present at the recurrence time, so that the aligned molecules can be studied in their natural state. Collision experiments using aligned samples will provide new

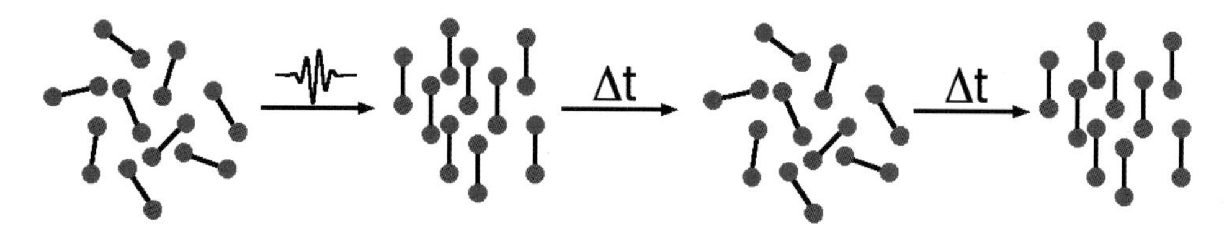

FIGURE 5-4 As described in the text, a random sample of diatomic molecules kicked by a short laser pulse will tend to align themselves right after the pulse. Although this alignment does not distinguish between the two ends of the molecule, methods have been developed to do so. The molecules continue to rotate at different rates and after a few picoseconds dephase into a randomly oriented collection of molecules. However, in the absence of other effects, after a time (tens to hundreds of picoseconds) the molecules will rephase to produce the aligned sample again.

information on the anisotropy of molecular interactions, yielding insight into how reactions actually occur. In addition, the ability to align single molecules will dramatically simplify the analysis of single-shot, and even single-molecule, structure determinations that may become feasible at new x-ray light sources like the Linear Coherent Light Source, currently under construction.

LOOKING TO THE FUTURE: CAN WE SEE AN ELECTRON'S MOTION?

When we peer at an atom or molecule using a femtosecond pulse, the motion of the lighter, rapidly moving electrons is lost in a blur. If we could develop sufficiently short pulses to allow us to strobe the motion of electrons themselves, we would truly enter a new and astonishing realm of science. Two approaches are possible: slowing down the electrons or speeding up the strobes.

Slowing Down the Electrons: Rydberg Electrons

Much insight into the behavior of the electrons comes from the study of a highly excited species known as Rydberg atoms. In these atoms, one or more electrons is excited to a very-high-energy orbit, with just less than the energy necessary for ionization—that is, the energy necessary to escape the influence of the charged core of the atom. This Rydberg electron spends most of its time moving very slowly compared to the electrons in the atomic core: The period of a typical Rydberg electron in an orbit with a radius of ~100 nanometers is 5 picoseconds—200,000 times longer than the 25-as period of an electron in the ground state of a hydrogen atom. Such slow electrons can be filmed in pump-probe experiments using commercially available femtosecond lasers. Studies of Rydberg atoms and molecules have provided enormous insight into the motion of the electrons, and many of their results can be extrapolated to the behavior of the core electrons. For example, time-resolved studies of doubly excited Rydberg atoms—in which two electrons are pumped into highly excited orbits—show how the motions of the two electrons are correlated, as well as how their collisions result in the ejection of an electron or a photon from the atom. Such correlated electron motion drives many of the important processes in chemistry and biology and is often a key factor determining the character of novel complex materials.

SPEEDING UP THE PULSE: ATTOSECOND SCIENCE

The electrons in most atoms and molecules do not move on such a convenient timescale. Yet there is much to be learned from this motion. When sunlight reacts in our skin to help us manufacture vitamin D, how exactly do the electrons permit the carbon ring to open? Where do the electrons move during this process? Or in chemical physics, can we learn how to make organic solar cells for energy applications? How do x rays damage DNA, and can we learn how to use and control this process for applications in medicine?

Making Attosecond Pulses

To answer questions like these we must find ways to see electron motion directly, and this means that we must shorten the time duration of the strobe. This has led to the birth of "attoscience," the creation and use of laser pulses less than 1 fs long. To understand this achievement, it helps to realize that green light, for example, has an electric field that oscillates with a period of 1.5 fs. The cycle period becomes shorter as the color moves towards blue, corresponding to higher energy photons. A pulse of light must be longer than one cycle, so attosecond pulses are in the ultraviolet or x-ray region. These attosecond pulses are generated by converting pulses of visible and infrared laser light through a process known as high-harmonic generation, described in Chapter 4. By focusing a high-intensity femtosecond laser into a gas, the laser pulse literally rips the most loosely bound electron from the atom and then smashes it back into the atom. Through this process, a tightly collimated beam of ultraviolet or x-ray radiation is generated. Under the right conditions, these x-ray beams can have durations of a few hundred attoseconds, much shorter then the original laser beams, and can also be directed and laserlike.

Using Attosecond Pulses

Just as important as generating the attosecond pulses are the means to propagate, characterize, and use them productively. New methods and techniques have already been developed to address each of these issues. Box 5-1 showed how these tools have been used together to probe the decay of highly energized atoms of krypton. In the future, we may be able to record the motion of electrons as they undergo transitions between atomic orbitals and even follow the correlated dynamics of multielectron motion in molecules. With attosecond techniques, scientists could capture and manipulate the electrons that serve as the glue that holds atoms and molecules together. This would be a regime where the motion and the interactions of this cloud of electrons swarming around an atom or a molecule can be observed and controlled in real time. To date, simple experiments have observed and manipulated the dynamics of this electronic cloud. In the future, we may be able to follow the correlated motion of entangled multielectron wave packets to understand and to control processes that cannot be accessed using any other type of probe.

HARD PHOTONS AND FAST ELECTRONS

Attosecond science confronts directly the complementarity between photon energy and pulse length: The shorter the pulse, the higher energy the photon must be. In turn, the higher energy the photon, the more it involves interactions with the fastest (innermost) electrons of the atom. Performing new experiments in attosecond physics requires considerable knowledge of the physics of inner-shell phenomena, and one might ask where this information comes from if the time-resolved techniques are only now becoming possible. Much of this information comes from experiments that focus not on time resolution but on energy resolution. For example, experiments can be performed to measure the precise energies of excited states of atoms and molecules and their ions, or to measure the energies and angular distributions of the electrons and other fragments produced when these excited states decay. Experiments can even be performed that determine the spin of the electrons. Synchrotron light sources such as the Advanced Light Source, the Advanced Photon Source, the National Synchrotron Light Source, and the Stanford Synchrotron Radiation Laboratory have enabled such experiments by providing intense sources of far-ultraviolet light and x rays, albeit with relatively long pulse durations. Experiments at these facilities have substantially improved our understanding of complex processes in the inner shells of atoms and molecules and enhanced our ability to characterize materials. This knowledge will continue to grow and is likely to prove invaluable in the development of attosecond science. While the present chapter highlights opportunities in time-domain studies of atoms and molecules, dramatic progress is also being made from the complementary perspective of energy-resolved processes, as described in Box 5-3.

IN REAL LIFE, TIMESCALES OVERLAP

A discussion on the difficulty of quantum calculations for all but the simplest molecules appears earlier in this report. What progress has been made in theory is due to approximations, none more important than the separation of timescales. As was learned above, different types of atomic and molecular motion have characteristic timescales. As a result, it is often possible to separate a complex problem into simplified parts: The fast motion of electrons can be studied independently by assuming that the vibrational and rotational motions of the atoms in a molecule are frozen on the relevant timescales, while slower motions like rotations can be studied by averaging over the much faster electronic and vibrational motions. Such separation of timescales is used as the basis for theoretical approximations that simplify the calculation of molecular properties and dynamics and provide a conceptual basis for understanding the fundamental physics. Nevertheless, many

BOX 5-3
Coincidence Measurements

In the study of energy-resolved processes, relatively long pulses of light are necessary to define the energies precisely, and such sources of light in the far-ultraviolet and x-ray regions have become available at synchrotron radiation facilities or using harmonics of high-resolution lasers. Because these sources are so much more intense than previously available sources, experiments that were once only a dream have now become routine. Perhaps the most revealing experiments employing these intense tunable light sources are enabled by the concurrent development of new imaging detectors and coincidence techniques, which allow the simultaneous measurement of multiple parameters associated with the fragmentation of the target atoms and molecules. For example, the absorption of a single x-ray photon by an atom often results in the ejection of two or more electrons. The measurement of the energies of these electrons, along with the angular distribution of their velocities with respect to each other and the polarization of the light, can provide enormous insight into the physics of this process, particularly when determined as a function of the photon energy. Inner-shell absorption in molecules can also produce multiply charged ions, and these often fragment into two or more ions. In a diatomic molecule undergoing rapid dissociation, the detection of an ion at a particular angle serves to fix the molecular axis at the moment of ionization, and the coincident detection of the photoelectron energy and direction provides the photoelectron angular distribution in the molecular reference frame (Figure 5-3-1). Such experiments provide an alternative to laser-alignment techniques for recording the spectra of fixed-in-space molecules. These multiply differential measurements, in which the ionization and dissociation dynamics are measured over many dimensions of parameter space, are often essential in differentiating among several possible decay mechanisms and have produced a new appreciation of the limitations of standard theoretical approximations, as well as an improved understanding of correlated electron motion and its role in ionization and dissociation mechanisms.

important phenomena occur when these approximations break down, providing a great challenge for theorists to develop new approaches and approximations valid in these new regimes.

The separation of timescales for molecular vibration and electron motion results from the vastly different masses of the atomic nuclei and the electrons: The heavier nuclei generally move much more slowly than the lighter electrons. Shortly after the birth of quantum mechanics, this realization led to the development of the Born-Oppenheimer approximation, in which it is assumed that the electron motion within a molecule can follow instantaneously any changes in the positions of the nuclei. This approximation allows calculating the potential energy of the molecule by freezing the geometry and solving the electronic problem.

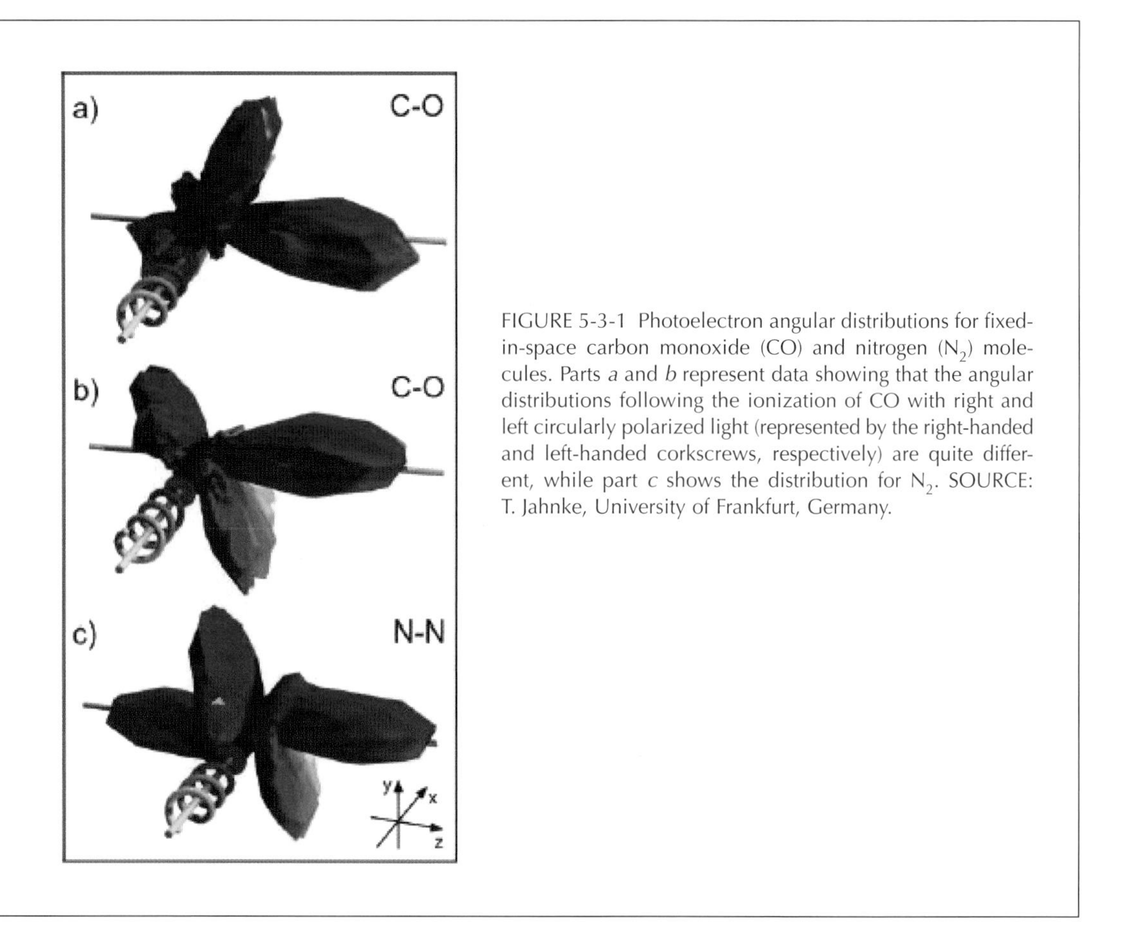

FIGURE 5-3-1 Photoelectron angular distributions for fixed-in-space carbon monoxide (CO) and nitrogen (N_2) molecules. Parts *a* and *b* represent data showing that the angular distributions following the ionization of CO with right and left circularly polarized light (represented by the right-handed and left-handed corkscrews, respectively) are quite different, while part *c* shows the distribution for N_2. SOURCE: T. Jahnke, University of Frankfurt, Germany.

Important situations arise, however, when the usual separation of timescales breaks down and the electronic and vibrational motion cannot be treated independently. One way in which the Born-Oppenheimer approximation can break down is if the potential energy surfaces of two different electronic states of the molecule cross—that is, if they have the same energy for some geometry of the molecule. If the two states interact, very small changes in the geometry of the molecule can produce large changes in the motion of the electrons (i.e., the electronic structure) as the molecule flips between states. As a result, the electrons cannot instantaneously adapt to the vibrational motion, and the Born-Oppenheimer approximation breaks down. The character and behavior of the molecule can change dramatically as it switches from one surface to the other. In polyatomic molecules, the intersection of multidimensional potential energy surfaces can lead to intriguing structures such as conical intersections, which play a critical role in how energy deposited into an isolated molecule is redistributed among the electronic and vibrational motions of the molecule and, ultimately, in how the molecule reacts or decomposes. For example, it was only when these nonadiabatic interactions were carefully taken into consideration that one of the simplest chemical reactions—the dissociative recombination of H_3^+ with an electron—could be properly modeled theoretically. Indeed, nonadiabatic transitions play an important role in many fundamental processes in physics, chemistry, and biology, including photochemistry and nonradiative relaxation, charge transfer and photosynthesis, and solvent caging effects in liquids. There are some environments in which nonadiabatic interactions are inescapable. For example, for particles a few nanometers in size, the electronic-level spacings and vibrational energy spacings are comparable, suggesting that nanoscience will abound with nonadiabatic effects.

Here the ability to sculpt attosecond pulses and control the electronic motion within molecules could prove invaluable. The ability to control the electronic wavefunction will allow scientists to drive molecules through such surface crossings along different trajectories and follow their outcome, providing the means to map out the detailed character of the potential energy surfaces and the electron motion near the crossing and to elucidate mechanisms for the nonadiabatic processes. In such experiments, the motion of the nuclei could be monitored using ultrafast, high-energy x rays, such as those that will be produced at XFELs, discussed in Chapter 4. The advantage of using such x rays is that they excite the innermost electrons of the atoms, which provide the most precise definition of the atomic positions. Together, such capabilities will significantly advance the experimental study of reaction dynamics in molecular physics, chemistry, and biology and provide important clues to understand these dynamics.

Controlling the Ultimate in Timescales

The shortest pulses and most intense electromagnetic fields that have been applied to atoms and molecules are not generated by lasers at all. Rather, they have been obtained by accelerating charged particles to extremely high velocities and crashing them through a sample of the target atoms or molecules. In many ways, the effect of the sharp pulse of charge passing through the sample is similar to an extremely short laser pulse. For decades, the field of atomic collision physics has yielded insights into the response of atoms and molecules to short pulses of electromagnetic radiation. The field of atomic collision physics is sufficiently mature, and its record of accomplishments is so well established in the literature and by applications, that it is largely left to stand on its own powerful historical record in this report. The generation of high harmonics from short laser pulses is itself a collision phenomenon, involving the recollision of an electron with the parent ion left by the initial laser ionization.

Advances in laser technology allow us to ask new questions about collision physics. Can the element of control that the laser offers be extended to, and united with, this alternative approach to the production of ultrafast pulses? More specifically, can we control the outcome of an impulsive collisional encounter by imposing on the interaction region a laser pulse of sufficient strength and appropriate timing? With the current availability of laser pulses with field strengths that begin to approach those provided in collisions, we can alter the time-dependent fields felt by the colliding components in such a way that the motion of the electrons is changed. For example, theoretical calculations indicate that the likelihood of the transfer of an electron from one colliding atom to another in a moderate-velocity collision can be enhanced by an order of magnitude by applying to the target atom electric fields much weaker than the electric field provided by the atom's collision partner. The corresponding experiments, while challenging to carry out, are under way in several laboratories around the world. Electron transfer in the interaction between colliding systems plays a key role in a wide range of chemical and biological reactions, and the ability to control this process could have far-reaching consequences. Extensions of this conceptual approach include pump-probe experiments in which a laser-prepared (excited, aligned) target is bombarded by a fast ion, as well as heavy-particle diffraction images of molecules and thin solids using picosecond pulses of high-energy, heavy-particle beams.

PROBING TIME-DEPENDENT MOLECULAR STRUCTURE WITH ELECTRONS

Imaging the nuclei in atoms and molecules with matter waves of subatomic particles has its roots in the earliest days of quantum mechanics. Electron microscopes are now used to examine everything from nanostructures in cells to molecules on surfaces, to the inner structure of nuclei and baryons. Until now, however, the pictures have been long-exposure stills, not action-packed movies. Rapid developments of the last decade have added a new element to the mix: time-resolved electron diffraction with pulses only a few hundred femtoseconds in duration. By using femtosecond lasers to generate short electron bursts and high acceleration voltages and small currents to minimize the spreading of the electron bunches, time-resolved electron diffraction can now be used to make molecular movies with a time resolution as short as 500 fs. Beautiful pictures of the evolution of melting in solids and the twisting or isomerization of large molecules have been

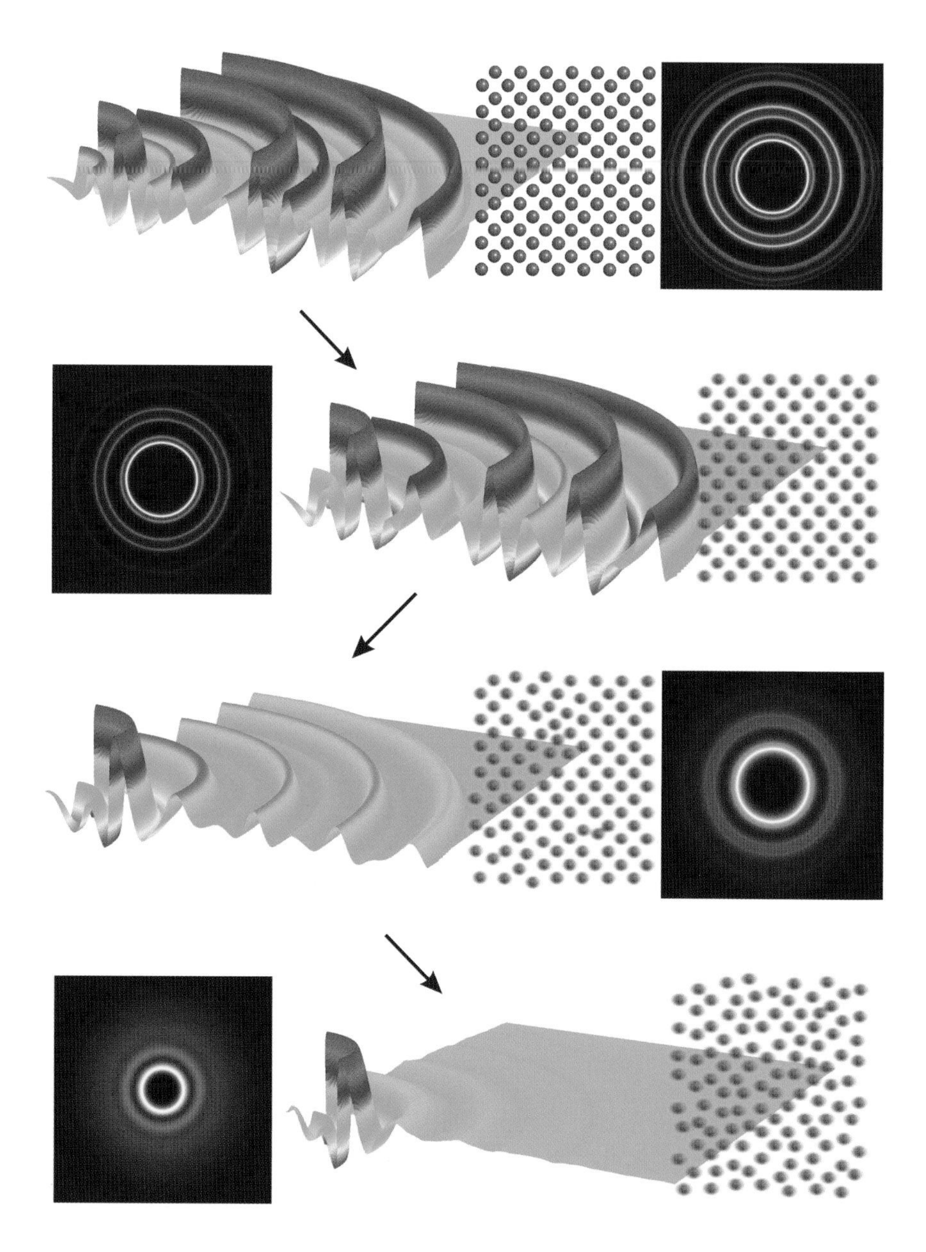

FIGURE 5-5 Ultrafast electron diffraction images, after Fourier filtering high-frequency noise, showing the melting of aluminum, captured in several stages only a few picoseconds apart. Each line shows the diffraction data and the reconstructed image of the aluminum lattice at a particular time as the aluminum melts. Femtosecond time resolution (600 fs) enabled the experiment to determine that the phase transition is thermally driven homogeneous nucleation as opposed to nonthermal electronic mechanisms. SOURCE: R.J. Dwayne Miller, B.J. Siwich, University of Toronto, Canada.

recorded (see Figure 5-5). The faster motions of typical molecular vibrations are still a blur, but the push to shorter timescales is on.

An In Situ Approach to Ultrafast Electron Scattering

In the time-resolved diffraction experiments discussed above, the source of the electron bunches and the sample to be studied are spatially separated. One way to improve the time resolution of the diffraction experiment is to probe the molecule with its own electrons (see Figure 5-6). Here, the ability to control the electronic wavefunction will be particularly valuable. As in the case of atoms, if an intense femtosecond laser is focused onto a molecule, an electron is ripped off the molecule and then slammed back into it. The electron wave packet becomes a short-wavelength replica of the longer-wavelength photon packet and becomes the probe with which the atom or molecule is interrogated. For example, when the returning

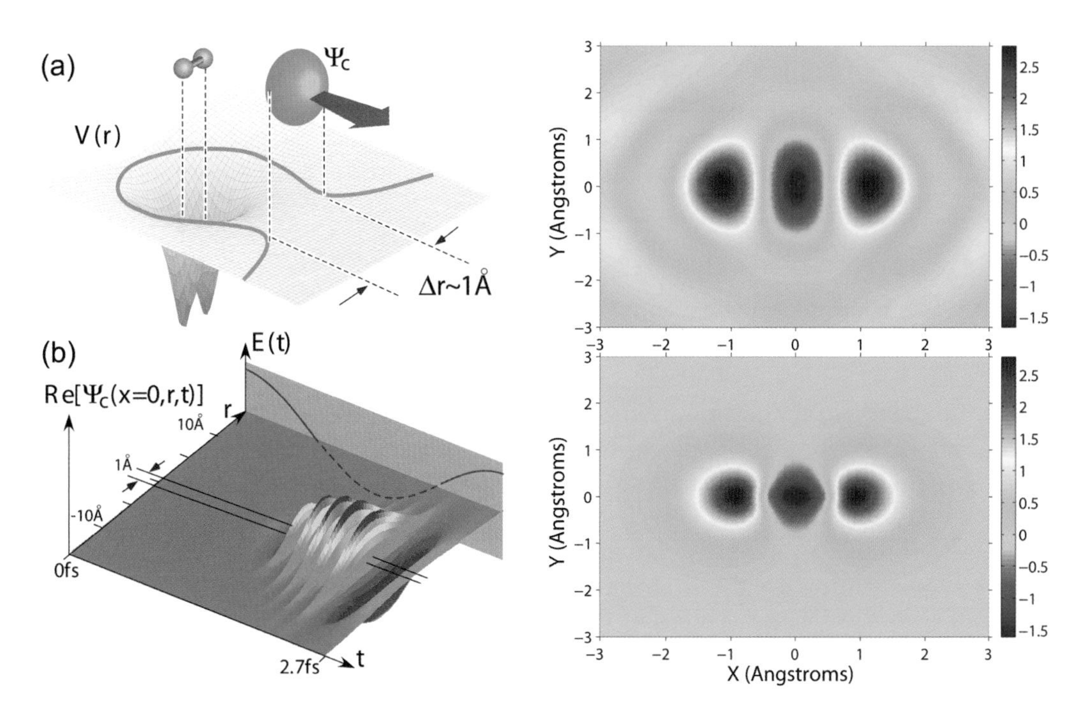

FIGURE 5-6 Molecular tomography. Illustration of an electron in molecular nitrogen as it experiences strong field ionization (*top left*) and recollision (*bottom left*) under the driving force of an intense laser beam. The VUV radiation produced in the recombination of the electron with its parent molecule is collected and analyzed to produce a snapshot image of the molecule (*top right*), which can be compared to a calculation (*bottom right*). SOURCE: D. Villeneuve, National Research Council of Canada.

electron is captured by the parent ion, the high harmonic light so generated can provide information on the electronic structure of the molecule (see Figure 5-6). In addition, the rescattering electron itself can be diffracted by the molecule and thus also carries detailed information on the molecular structure. Because the recollision is nearly instantaneous—the time interval between each ionization and recollision event is only a fraction of a cycle of visible light, or approximately 3 fs—the electron diffracts from a relatively unchanged molecule, since the positions of the atomic nuclei cannot change appreciably on this timescale. Can illuminating a molecule with its own electron be useful as a probe of the dynamics of the molecule itself? Absolutely! Because the electron wave packet is emitted from, and immediately refocused onto, the molecule to be interrogated, there is little time for the electron wave packet to spread. As a result, the effective intensity of the returning electron wave packet is immense (easily exceeding 100 billion A/cm^2). No macroscopic source of electron beams can approach this value. Thus far, theoretical studies of the self-electron-diffraction technique have shown that it is possible to study the motion of the nuclei in diatomic molecules. The future will surely see studies on more complex molecules, and even on molecular reactions. Furthermore, by shaping the laser pulse in time, it will be possible to sculpt the emitted electron wave packet in both time and space to optimize the characteristics of the returning wave. Such advances in the control of electron motion could lead to a new age of electron diffraction and significantly enhance efforts to characterize the structure of complex molecules.

THE FUTURE

As our sophistication with these new experiments grows and new techniques emerge, important questions will arise regarding the use of shaped pulses as reagents in chemical and other processes. These questions will involve the speed and uniqueness of the optimization techniques, the degree of control that is achievable, and the cost of the overall approach relative to other methods. Big questions also remain at the fundamental level. For example, we must ask why a particular pulse shape is optimal. One approach to this question is to take the optimized pulse apart and, through a combination of more traditional pump-probe experiments and detailed quantum chemical calculations, to reconstruct the effect of the pulse on the molecule in a series of steps to see how the result is achieved. While such reverse engineering can help elucidate the control mechanism, it is clear that new insight and modes of understanding are also necessary to make the most of the results.

How fast can we make physical devices? Not faster than the atoms and molecules themselves operate, but perhaps just as fast? It is thus not difficult to see the importance of continuing to press our ability to observe and control matter to this timescale. Is it a stretch to say that in a decade of two we will see computer cycle times approaching the timescales of electrons in individual molecules? Or that designer molecules for health care will be created and reproduced using molecule-milling machines in which every atom is placed individually? Or that combustion reactions will be made entirely by-product-free by sending the reactants through aligning and exciting laser-preparation instruments on their way to the burn? Perhaps, but then again much of the technology we take for granted today would have seemed completely out of reach 20 years ago. At the heart of this progress in technology is the constant search for faster observation and control. We stand at the threshold of the age when the timescale at hand is in some sense ultimate, namely that of the atoms and molecules which themselves make up matter. Control at this timescale is certain to launch truly revolutionary technology.

6

Photonics and the Nanoworld

How will we control and exploit the nanoworld? The scale of the nanoworld is smaller than a wavelength of visible light. This places it between the classical world of microscopic objects such as living cells and the quantum world of atoms and molecules. Nanoscale structures are rich in promise for novel AMO research because they have nonintuitive but useful physical, chemical, and biological properties that come from their submicroscopic size. These tiny structures present unique opportunities to tailor material properties for a tremendous range of potential applications. In optical physics, nanoscience promises efficient optical switches, light sources, and photoelectric power generators. In the next decade, new nanomaterials may dramatically improve optical microscopes or reduce the feature size in semiconductor chip fabrication. Other applications include precisely periodic nanostructured materials or photonic crystals, single-photon sources and detectors, nanostructured electron emitters for flat-panel displays and television screens, sensors for environmental monitoring, and biomedical optics, with applications such as killing cancerous cells via localized optical absorption and heating.

The nanoworld exists at the interface between the classical and quantum worlds, providing the opportunity to explore the transition between these worlds, as well as the prospect of using features from both worlds to produce exciting new behaviors and technology. Nanoscience, the study of the nanoworld, is an inherently multidisciplinary field of research that requires concepts, methodologies, and tools from a broad range of scientific disciplines. AMO physics is a touchstone for understanding many of the new phenomena of nanoscience and has provided numerous tools for performing forefront research. In return, nanoscience promises to provide fundamental and technological breakthroughs that will benefit both the scientific community and society at large. AMO physics is already benefiting from new tools that are enabled by breakthroughs in nanofabrication and nanoengineering. This chapter highlights specific opportunities that lie at the interface between nanoscience and AMO physics, with a particular emphasis on photonics.

The allure of nanoscience can be understood by considering what happens if a macroscopic gold nugget is divided into ever smaller pieces.[1] At first, the piece of gold will retain all of its typical characteristics—it simply gets smaller. When the pieces reach a few micrometers in size, we are no longer able to distinguish individual particles with our eyes. However, given enough of them, we still see gold dust. However, when the particles reach a size of ~100 nm, something very dramatic happens: The particles change color. For particle diameters between approximately 100 and 30 nm (i.e., for particles containing between approximately 30 million and 1 million gold atoms) the particles change from red or yellow, to green or blue. The particle's color is determined by its size. Quite amazingly, these colored gold particles have been known since the Middle Ages, when they were used to make beautiful colors in stained glass windows. Of course, the medieval artisans did not know that they were using nanotechnology, or even why the gold produced the colors it did: They just knew that a particular process produced a beautiful effect (see Figure 6-1).

It is only in the last few years that we have begun to understand the size-dependent changes that occur in gold and other metallic nanoparticles. The size of a nanoparticle determines the character of its surface plasmons, a type of collective motion of the electrons within the particle that gives rise to its color. The strong dependence of the particle's characteristics (in this case its color) on the size of the particle is one of the key features of nanoscience. With our understanding of the nature of the color changes comes the opportunity to tune the particles to achieve the behavior we desire.

OPPORTUNITIES IN SIZE-DEPENDENT DESIGN

The nanoworld is intriguing because it is not just the color of nanoparticles that depends on size. At the nanoscale, a wide range of physical, chemical, and even biological properties can be strongly dependent on the particle size. We expect that by working at this scale, we will soon be able to design and engineer structures with a tremendous variety of desirable features. This is exactly what nature does in the inner workings of every cell. Proteins are nanoscale molecules that are assembled by nanomachines called ribosomes, which connect a series of amino acid molecules together according to a pattern furnished by RNA, another nanoscale molecule. Each protein is essentially a nanoengineered molecule that has been optimized to perform specific cellular functions. A typical cell is filled with numerous nanomachines that serve a multitude of functions, including regulation

[1]For more information, see Mark Ratner, *Introduction to Nanoscale Materials Behavior—Why All the Fuss?* Available at <http://www.blueskybroadcast.com/Client/ARVO/trans/arvo_ratner_RF_OK.pdf>.

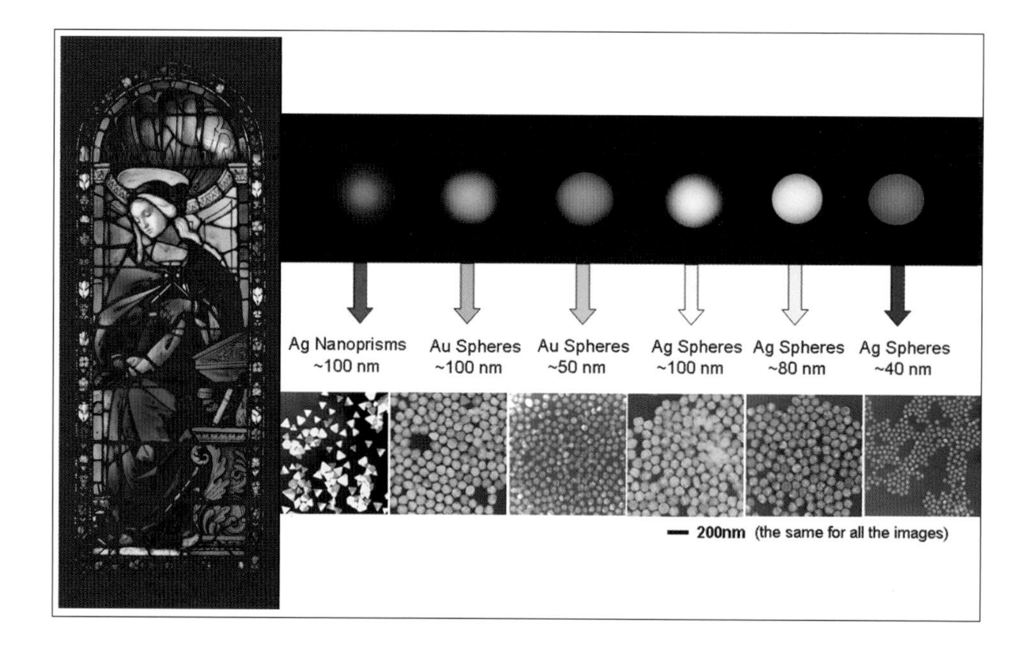

FIGURE 6-1 *Left:* Gold and silver nanoparticles are responsible for some of the beautiful colors in stained glass windows from medieval times. SOURCE: National Gallery of Art, Washington, D.C. *Right:* The colors of the gold and silver nanoparticles vary because of their different size, shape, and composition. SOURCE: Chad A. Mirkin, Institute for Nanotechmology, Northwestern University. These figures were taken from a slide in a presentation entitled "The Shifting Plate Tectonics of Science" by Arden L. Bement, Jr., Director, National Science Foundation, at the American Ceramic Society, Baltimore, Maryland, April 10, 2005, available at <http://www.nsf.gov/news/speeches/bement/05/alb050410_ceramicsociety.jsp>.

of the passage of food and waste through the cell membrane, generation of fuel for cellular functions, recognition of invading viruses, and full replication of the cell itself. Indeed, it has been said that the nanoscale is the design scale of nature. The amazing power of cellular machines has long provided inspiration to scientists. With the growth of nanoscience, we are in a position to develop and implement an extraordinarily broad range of new bioinspired materials, catalysts, and machines. One of the most exciting of these is the possibility of creating new power sources based on the light-harvesting photosynthesis machinery of plants.

The recent explosion of interest in nanoscience and nanotechnology has been fueled by the development of a host of new capabilities for visualizing, constructing, and manipulating matter at the nanoscale. Broadly speaking, nanoscience and nanotechnology involve three basic themes:

- Research and development at the atomic, molecular, and macromolecular level involving length scales of approximately 1 to 100 nm (approximately 50 atoms to 50 million atoms);
- The creation and use of structures, devices, and systems that have novel properties and functions as a result of their size; and
- The ability to control or manipulate light and matter on the atomic scale.

The committee expects that in the coming decade, nanoscience will lead to tremendous advances for both science and society. In what follows, it describes some of the rapidly evolving approaches to visualize, manipulate, and ultimately harness the nanoworld that are enabled by AMO physics.

VISUALIZING THE NANOWORLD

The construction of our eyes allows us to see visible light or, more generally, electromagnetic radiation with wavelengths between approximately 400 nm (violet) and 750 nm (red). The wavelength of the light limits the ultimate size of an object that we can directly see using conventional optics. Thus, even with the best optical microscopes, it is not possible to resolve objects less than approximately 400 nm apart, because the wavelength of the light is longer than the separation, and the two objects therefore appear blurred together. If this is the case, how can we visualize the nanoworld? In this section, three different approaches to answering this question are discussed, each of which involves tools from AMO physics: reducing the wavelength, using scanning probe microscopies, and exploiting nanotechnology to build a better microscope.

Reducing the Wavelength

One route to improved resolution is to reduce the wavelength of the light used for viewing. Shorter wavelength light, corresponding to the ultraviolet and x-ray regions of the electromagnetic spectrum, provides better resolution than conventional optical microscopes. To date, soft x-ray microscopes have achieved a resolution of a few nanometers. In principle, shorter wavelength, harder x rays should allow even better resolution, but to date such microscopes have been limited to a resolution of ~20 nm. This resolution is limited not by the wavelength but rather by the ability to fabricate the appropriate x-ray lens (known as a zone plate) that is used to focus the x rays. Advances in nanofabrication techniques will one day override this limitation, allowing x-ray microscopes with subnanometer resolution to be built.

Often objects viewed with ultraviolet light or x rays look much different than when they are viewed with visible light. In general, visible light interacts with the most loosely bound electrons in molecules, while x rays interact with the most strongly bound electrons. The former tend to be delocalized over the molecule, while the latter tend to be localized on individual atoms. By illuminating the sample with particular wavelengths of light, or by detecting fluorescence at particular wavelengths, it is possible to use different techniques to map out the spatial distribution of different elements in the sample. For example, it is possible using a hard x-ray microscope to map out the spatial distributions of different elements within a single cell. Other techniques can be used to provide even more information. For example, nonlinear optical techniques such as multiphoton excitation and Raman spectroscopy are now being used to image specific chemicals, again with subcellular resolution, and to study time-dependent processes involving these chemical species. Although these advances are expanding rapidly into an increasing number of scientific disciplines and technological applications, it is particularly noteworthy that AMO science provides essential underpinnings for understanding the new contrast mechanisms and optical techniques.

Beyond x rays, can the resolution be pushed even higher? The quantum mechanical nature of matter gives both particle and wave character to light and matter. The heavier a particle and the higher its kinetic energy, the smaller its de Broglie wavelength. Electron microscopes use high-energy electron beams to view the sample: In this case the de Broglie wavelength is so short that it no longer limits the resolution. The appearance of samples viewed with an electron microscope is determined by the details of the electron-matter interactions, which have long been a subject of great interest for AMO physics. Using even heavier particles—namely, whole atoms—researchers in atom optics are working to create a new, high-resolution atom microscope with an ultrashort de Broglie wavelength.

Scanning Probe Microscopes

One approach to imaging the nanoworld is to dispense with the idea of "seeing" altogether and to use an alternative means of visualization. As an example, consider having fingers only a nanometer in diameter and slowly running them across some molecules sitting on a surface. The displacement of these fingers as they cross the molecules would be similar to that of the tip of an atomic force microscope (AFM), which allows one to create a nanoscale map of the surface roughness by measuring the deflection of an ultrafine tip on a cantilever as it is "dragged" across the surface. An example is shown in Figure 6-2. The AFM is but one of a wide range of scanning probe microscopies, each of which relies on a different molecular-scale effect for its contrast mechanism. The intermolecular forces and tunneling phenomena that provide the contrast mechanisms for these microscopies can best be understood using concepts from AMO physics.

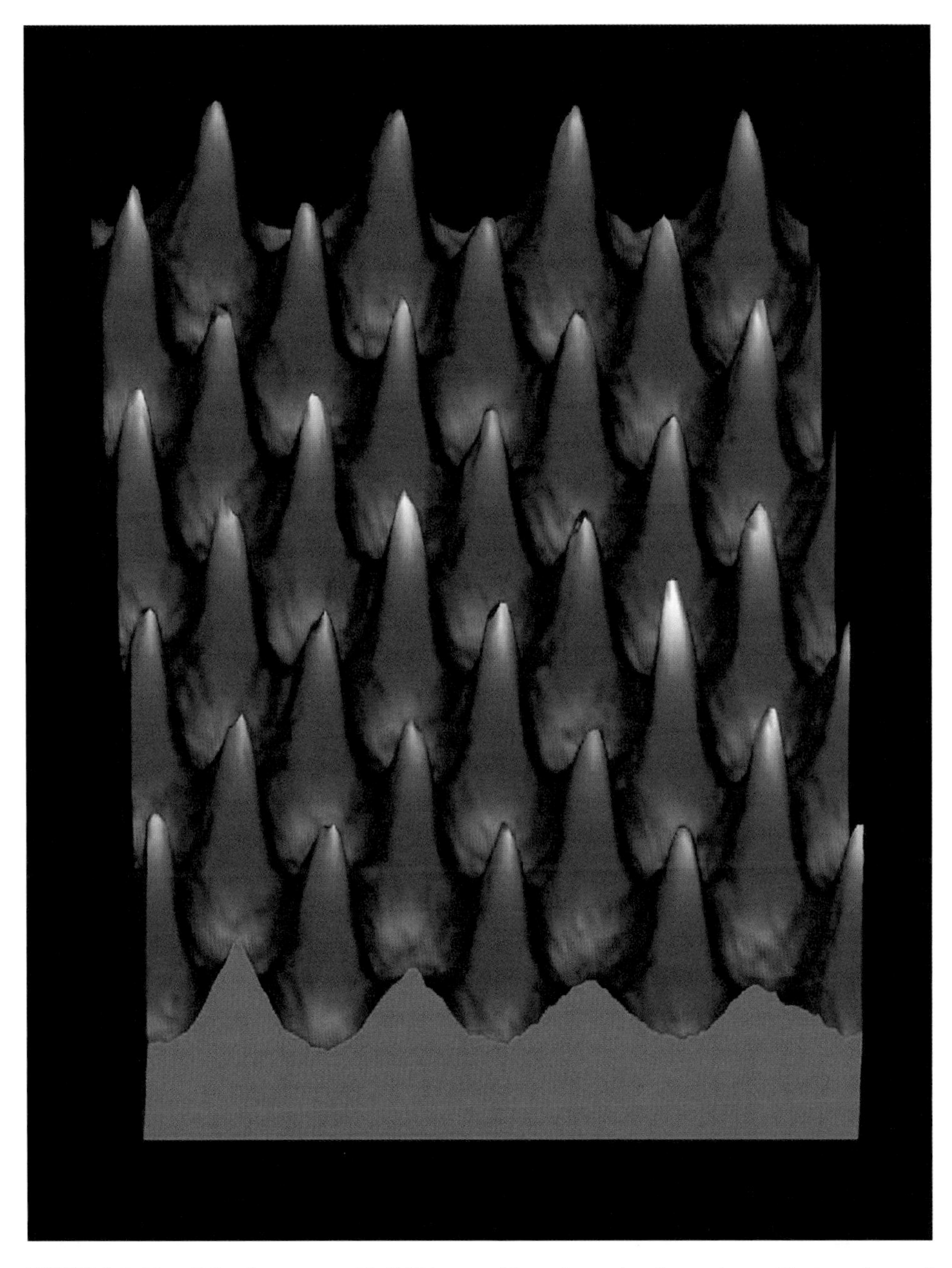

FIGURE 6-2 Visualizing the nanoworld. AFM image of two-dimensional array formed by laser-focused atomic deposition of Cr. Features are 13 nm high and 80 nm wide and are on a 213 × 213 nm square grid. SOURCE: J. McClelland, National Institute of Standards and Technology.

The scanning tunneling microscope (STM) is a particularly powerful tool that relies on the measurement of tiny electrical currents that pass between the tip and the surface, with changes in these currents reflecting the local densities of electrons on the surface. Such STMs have been used to map out the positions of individual atoms on the surface of a crystal, as well as the positions of atoms within a molecule. Perhaps even more exciting, the tips of these STMs have been used to move and position individual atoms sitting on a crystal surface, allowing the creation of controlled structures with atomic dimensions.

An optical version of these scanning probe microscopies also exists—namely, near-field scanning optical microscopy (NSOM). This technique provides subwavelength resolution that is well below the conventional limit of optical microscopy. In one form of NSOM, the tip used in an AFM or STM is replaced by an optical fiber tapered to a tip approximately 50 nm across; this tip is positioned a few nanometers above the surface of interest and then scanned across it. Laser light is sent down the fiber and tunnels out of the tip, illuminating the sample over an area approximately equal to the diameter of the tip. Scattered light or fluorescence from the sample is then detected, allowing an optical image with approximately 50-nm resolution when a laser wavelength of 500 nm is used. This approach has also been pursued using ultrafast lasers, allowing the study of time-dependent phenomena with unprecedented spatial resolution. Many additional new techniques, such as multiphoton microscopy and ultrafast confocal microscopy, are now being developed to apply optical methods to probe nanoscale and even single-molecule processes in exquisite detail.

Using New Materials to Build a Better Microscope

Can nanoscience enable us to build a better microscope? One intriguing possibility is the development of a new generation of microscopes using properties of light that were previously inaccessible. Conventional lenses have what is known as a positive index of refraction. When light passes from one medium to another (e.g., from air to a quartz lens), the ray is bent by an angle that depends on the difference between the refractive indices of the two materials. These positive index lenses create images by capturing the propagating light waves emanating from an object and bending them into a focus. However, the resolution is limited to roughly one-half the wavelength of the illuminating light—that is, the diffraction limit. The electromagnetic field emanating from an object includes not only propagating waves but also near-field evanescent waves that decay exponentially as a function of distance away from the object. These evanescent waves carry the finest details of the object, but unfortunately they cannot be recovered by conventional positive-index lenses—and all naturally occurring materials have a positive index.

Theorists predicted that a material with a negative refractive index could capture and refocus these evanescent waves, and in 1968 it was proposed that a perfect lens, or "superlens," could be made from a negative-refractive-index material. In such a superlens, the electromagnetic waves reaching the surface of the negative-index lens excite surface plasmons, which enhance and recover the evanescent waves. Experiments have demonstrated the existence of materials with a negative index of refraction—called left-handed materials—for microwaves (see Figure 6-3). Researchers using a thin silver slab as a superlens were able to image objects as small as 40 nm, smaller by a factor of 10 than the limit of current optical microscopes. This important experimental advance not only resolved a controversial question about the nature of these negative-index materials but also opened the door to a variety of novel applications, including higher resolution optical imaging, nanolithography, and optical elements that exploit the evanescent waves of light.

CONSTRUCTING THE NANOWORLD

Synthesis at the nanoscale can proceed in two directions: starting on a larger scale and working down in size (top-down) or starting at the atomic scale and building up (bottom-up). A vast array of synthetic approaches using methods from physics, chemistry, materials science, and biology are rapidly being explored and developed, some of which combine aspects of both top-down and bottom-up design. What follows is only a brief overview of a few of the exciting possibilities for building structures in the nanoworld.

From the Top Down

One way to create nanostructures is to use optical lithography or laser ablation techniques to remove or add material to surfaces through a sequence of processing steps. The depth of removal or addition of material can readily be controlled at the level of single atomic layers. The width of features that can be written with these techniques is considerably larger. In addition to limiting the resolution achievable when viewing objects, the wavelength of light also places constraints on the resolution that is achievable when light is used to modify or write designs on surfaces using lithographic or ablation techniques. In particular, the resolution of the lithography that is required for manufacturing computer chips is limited by the wavelength of the light used. This is important because as the components on chips become ever smaller, new tools are essential to manufacture these chips. High-powered gas-discharge excimer lasers operating at ultraviolet wavelengths of ~193 and 157 nm are now commonly used in this process. AMO physics, in particular its insights into the electron-atom and electron-molecule collision processes in the

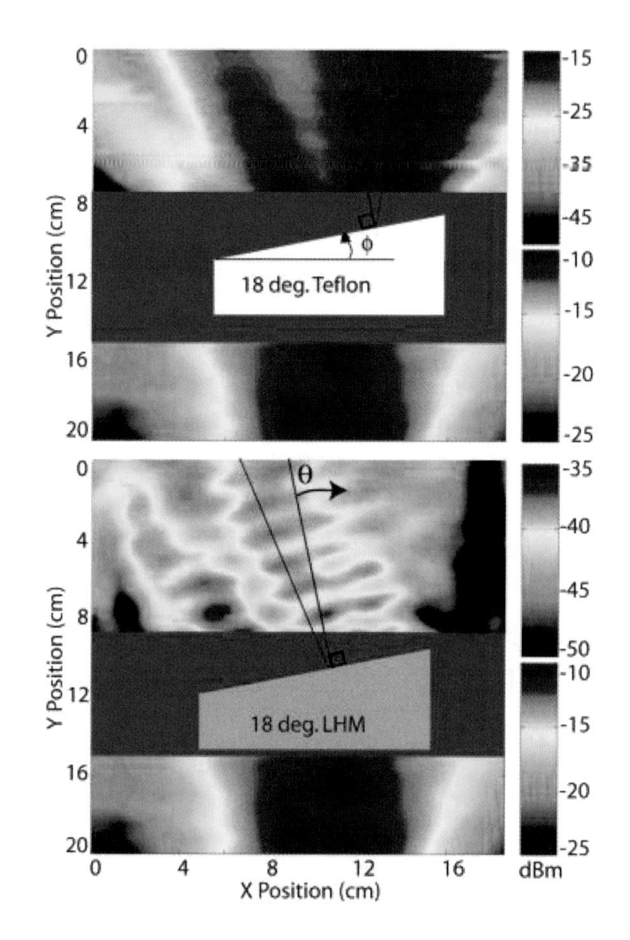

FIGURE 6-3 The terahertz frontier for negative-index materials. New sources and new applications of terahertz radiation have been forefront activities in AMO science over the past decade, and now this research is making connections to nanoscale materials. Many of the issues that arise when the size of the structure is comparable to the wavelength of the radiation have been understood and addressed for some time in studies using radiation at much longer wavelengths. For microwaves, the wavelengths are on the order of centimeters to millimeters, so that the construction of subwavelength structures is a considerably easier engineering task. As shown here, the first demonstration of a negative-index material was performed using microwave radiation. The top half of the figure shows microwave radiation passing through a wedge of a normal material (Teflon) and being refracted in a positive direction when it leaves the wedge. The lower half of the figure shows microwave radiation passing through a wedge of a negative-index material (left-handed material) constructed of wires and rings and being refracted in the opposite direction when it exits the material. In principle, materials with negative refractive indices at much shorter terahertz or even infrared wavelengths might be constructed using nanofabrication techniques. SOURCE: Andrew Houck, Yale University, reprinted figure with permission from *Physical Review Letters* 90, 137401 (2003). Copyright (2003) by the American Physical Society.

discharge and decay processes of the highly excited excimer molecules, played an essential role in the development of the high-power lasers that are now standard in the industry. The desire for still shorter wavelengths delivered from devices of reasonable size and efficiency has provided one of the strong motivations for tabletop x-ray sources.

Techniques based on scanning probe microscopies have also been used to modify surfaces at the nanoscale and may ultimately provide general-purpose techniques to do so. For example, light from a near-field scanning optical microscope tip can be used to induce photochemistry or to ablate materials from surfaces with nanoscale resolution. With new tools and techniques, our ability to modify materials at the nanoscale is rapidly growing. Even free atoms can now be manipulated at the nanoscale (see Figure 6-4), giving rise to the field of atom nano-optics, a subfield of the extensive domain of research involving the laser control of neutral atoms and molecules.

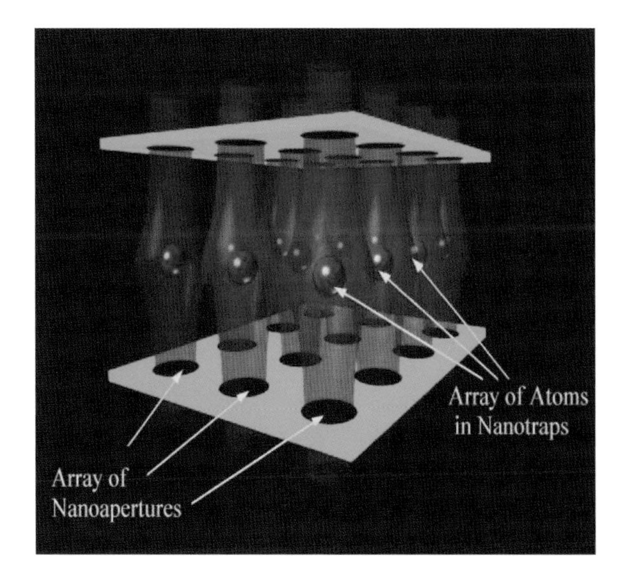

FIGURE 6-4 An array of atom traps. Light in the vicinity of holes and structures much smaller than its wavelength can have a number of interesting properties, allowing the creation of regions of high or low light intensity with sizes much smaller than this wavelength. With the appropriate structures (shown in green), these light fields can be arranged to produce a series of optical atom traps (atoms shown in blue) that are separated by distances much smaller than the wavelength of light. SOURCE: V. Balykin, V.V. Klimov, and V.S. Letokhov, Russian Academy of Sciences.

From the Bottom Up

For several decades, AMO physicists and physical chemists have studied how atomic and molecular properties change as one goes from a single atom or molecule to a bulk sample containing many millions of atoms or molecules. The study of clusters of carbon atoms revealed a special stability for the cluster containing 60 atoms. In this cluster, the atoms are arranged in the geometry of a miniature soccer ball only 0.7 nm in diameter. The discovery of this cluster, now known as buckminsterfullerene, was one of the significant events heralding the dawn of nanoscience, and the discoverers were awarded the Nobel prize in chemistry in 1996. The C_{60} molecule ("buckyball") was soon followed by other fullerenes containing many more atoms of carbon and taking different shapes. One of the most interesting and prominent of these is the buckytube, or carbon nanotube. Nanotubes generally consist of hexagonal lattices of carbon atoms arranged spirally to form concentric cylinders. Single-wall nanotubes have a typical diameter of approximately 1 nm, while multiwall nanotubes consist of between 2 and 30 concentric tubes with an overall diameter of 30-50 nm. Carbon nanotubes are usually near-perfect crystals and range in length from a few tens of nanometers to several microns. They have unique properties which, like those of diamond, arise from their nearly perfect structure. Depending upon the details of this structure, a nanotube can act either as a highly conductive metal wire or as a semiconductor. Nanotubes have been used to build the first room-temperature transistor ever made from a single molecule and are widely expected to be the key ingredient for nanoelectronics that will vastly extend the power and shrink the size of computers and other smart devices. They are also efficient electron emitters, which has led to their application in flat panel displays, as described in more detail below.

A second example is provided by the many studies of small water clusters that have been performed in an attempt to understand the character of liquid water and ice. By using positively or negatively charged water clusters, it is possible to select species with a specific number of water molecules and thus to study the transition from molecular to bulk behavior in a step-by-step manner. Recent studies using infrared spectroscopy have led to new revelations about how such clusters arrange themselves to accommodate a positive or negative charge. As methodologies improve, an increasing number of powerful techniques can be applied to these studies, dramatically increasing our understanding of liquids. A final example is the strong size dependence observed for the catalytic properties of metal clusters. New techniques are now being explored to deposit size-selected clusters on surfaces and to characterize this size-selective catalytic activity. One day, such techniques could be scaled up to provide new industrial catalysts with improved efficiencies and selectivities.

EXTENDING THE PROMISE OF THE NANOWORLD

Many other breakthroughs can be anticipated based on further developments in nanoscience and technology. For example, there could well be significant improvements in solar energy conversion due to nanostructured solar collectors and new photovoltaic materials. There will likely be advances leading to more energy-efficient lighting. Nanoscience will doubtless have a broad impact in materials science as well. Indeed, we are entering a decade where stronger and lighter materials with novel characteristics will continue to improve efficiency and fuel use in transportation. New catalysts will provide low-energy chemical pathways that will be used both in the chemical industry to produce cheaper products and in remediation and restoration projects to break down toxic substances in the environment. We can imagine a safer and healthier world with extremely sensitive chemical, biological, and industrial sensors. Nanoscale titania and zinc oxide particles are already used in sunscreen formulations with improved UV protection, increased transparency (to eliminate their chalky appearance), and antibacterial activity.

To date, the most concentrated effort to apply the tools of nanoscale synthesis, characterization, and modeling has been in studies of the electronic properties of semiconductors and related materials. This effort has been driven primarily by the vital importance of semiconductors to information technology. Nevertheless, it is quite clear from this experience that for a wide range of materials, many other novel combinations of properties can be expected to emerge in the nanoscale regime. Here the committee describes some applications in optics and photonics.

Controlling Light with Photonic Crystals

One fertile area of research exploits the concept of the photonic crystal. Photonic crystals are periodic nanostructures that are designed to control and manipulate the flow of light in one, two, or three dimensions using the principles of diffraction (see Figure 6-5). For the crystals to work effectively, the spacing within the periodic structure must be the same order of magnitude as the wavelength of light—in other words, visible light requires periodicity on the nanoscale. One application of this field is the creation of new optical devices that confine light using internal microstructures. The best known of these materials are "holey" fibers—microstructured fibers that trap and concentrate light very effectively. Potential applications of this technology include studies of cavity quantum electrodynamics and development of quantum optical devices with capabilities such as emitting a single photon on demand. Controlling the emission of single photons has been a priority goal for application in quantum encryption systems, as discussed in Chapter 7.

Because true single-photon sources are not yet available, today's commercial

FIGURE 6-5 The common blue morpho butterfly gets it iridescent blue color from diffraction off what is essentially a photonic crystal structure. SOURCE: The Butterfly Pavilion, Westminster, Colorado.

quantum cryptography systems rely on photons from attenuated laser pulses as an approximation of the single-photon state. However, the produced state has a nonvanishing probability of containing two or more photons per pulse, leaving such systems susceptible to eavesdropping through a beam splitter attack—this is what makes the need for advances in single-photon sources for quantum cryptography so important. The first electrically driven planar photonic-crystal laser was recently reported by a team at the Korean Advanced Institute of Science and Technology. This photonic crystal laser consisted of an array of holes passing through layers of InGaAsP multiple quantum wells (MQWs), which results in lasing in the near infrared at a threshold current about one hundred times lower than typical MQW lasers. This reduction in the threshold current suggests one approach to a zero-threshold laser or a source that delivers a single photon on demand. Another approach to making an all-optical router/buffer could take advantage of recent advances in all-optical switching in silicon photonic integrated ring resonator structures (see Box 6-1).

An interesting variation on the theme of controlling light with photonic crystals was recently demonstrated. In these experiments, two-dimensional arrays of 390-nm polystyrene nanospheres were created by trapping the particles with overlapping beams of light generated by scattering off a prism. These arrays show

BOX 6-1
All-Optical Switching in Silicon Photonic Integrated Ring Resonator Structures

Using electron-beam lithography and plasma reactive-ion etching, researchers have constructed an all-optical silicon switch consisting of a straight waveguide (light-guiding channel) adjacent to a circular waveguide, or ring resonator with 450-nm-wide by 250-nm-high rectangular cross sections and a ring diameter of 10 microns. A 250-nm gap exists at the closest separation between the straight waveguide and the ring resonator. Ordinarily, information-carrying light pulses traveling in the straight waveguide would couple into the ring resonator. However, if the circumference of the ring resonator is a multiple of the wavelength of the light pulses, then the ring is resonant and effectively blocks the light pulses from coupling into the ring resonator. By using a control optical pulse of a different wavelength and focusing it on the ring resonator near the 250-nm gap, one produces enough optical absorption and subsequent photocreation of electrons to modify the refractive index of the ring to modify the resonance. Now out of resonance the light pulses are coupled from the straight waveguide into the ring resonator. Switching speeds as fast as 450 ps have been reported using this approach, which is orders of magnitude faster than the competing technology of thermo-optic switches. With modifications of this structure, one can easily imagine storing an optical pulse in this silicon ring resonator and switching it out as needed to produce the all-optical router/buffer described using a "slow light" approach (Box 4-1).

many of the dynamical features of molecular crystals, such as surface diffusion, migration of defects, nucleation of phase transformations, and "Ostwald ripening"—where two arrays coalesce into one. Because the light can be used to control the positions of the nanospheres in the array, this new approach may prove to be a valuable method to assemble matter on the nanoscale.

Atomtronics

Atomtronics is a new technology enabled by nanoscience in which ultracold atoms are trapped and manipulated on or near a microelectronic chip. This atom-chip amalgamation of cold atom and optical technologies on semiconductor chips will be an emerging area of nanoscience and nanotechnology in the next decade. The success of microelectronics has demonstrated the enormous potential of miniaturization for turning basic physics into applications. Today, researchers are exploring further miniaturization to nanometer and even atomic scales. The tiny clouds of ultracold atoms trapped and suspended barely above a semiconductor chip surface can behave as coherent matter waves under the influence of the electromagnetic fields in the microchip circuitry. The chip fields generate tiny forces that can be used to guide and manipulate the matter wave. This presents opportunities for the construction of matter-wave interferometers on microchips, which may serve as sensitive probes for gravity, acceleration, rotation, and tiny magnetic forces, as well as analogs of transistors and other solid-state devices in the more distant future.

Nanotubes in Televisions

Success stories to date in the use of nanotechnology for real-world applications include the recent entry of carbon nanotubes in the flat-panel display market. A new prototype nanoemissive display (NED) based on carbon nanotubes may challenge existing technologies of liquid crystal displays (LCDs), plasma displays, and organic light-emitting diode (OLED) displays. The key to the new NED is the ability to grow carbon nanotubes directly onto the display's glass substrate. On a back plate only 3 mm behind each subpixel, a small structure containing about a thousand carbon nanotubes is arranged such that a properly applied voltage excites each nanotube, which in turn bombards the color phosphors with electrons. The new technology was recently demonstrated with full-color video, and it is estimated that a 42-inch NED would consume only 40 percent of the power of a similar-sized LCD.

Nanotechnology in Medicine

AMO researchers are merging state-of-the-art optical physics and nanoscience to make important strides in health and medicine. For example, treating cancer cells without destroying normal tissue has been a long-standing problem in oncology. While chemotherapy uses biochemical means to attack cancer cells selectively, it often destroys many faster-growing normal cells as well, leading to hair loss and more serious side effects. Researchers across the country are now attacking this problem using the unique properties of nanoscale materials.

One approach to the treatment of tumors involves the use of the same carbon nanotubes discussed above. Thousands of carbon nanotubes can easily fit into a typical cell, and they absorb near-infrared light that passes harmlessly through the rest of the cell. Electrons in the nanotubes become excited by the near-infrared light and release this excess energy in the form of heat. A solution of carbon nanotubes under near-infrared excitation can heat up to 70°C in as little as 2 minutes. When these nanotubes were placed inside the cells and irradiated with a near-infrared laser, the cells were quickly destroyed by the heat, while the light had no effect on the cells without nanotubes.

A second approach to cancer therapy involves the use of nanoshells: virus-sized, metal-coated glass shells whose optical properties can be tuned to absorb or scatter specific colors of light (see Figure 6-6). It has been discovered that when gold-coated nanoshells are injected into the bloodstream, they selectively accumulate in tumors. By varying the diameter of the silica core (~100 nm) and the thickness of the surrounding gold shell (~10 nm) of these particles, the peak optical absorption can be tuned to the near-infrared. Irradiation by a fiber-coupled diode laser can lead to local temperature increases of up to 50°C. Researchers have found that tumors in nanoshell-treated mice disappear 10 days after this treatment, while tumors in animals without nanoshells continued to grow rapidly, suggesting that such local heating is very effective in destroying cancerous tumors.

As an ever more rapidly aging world population presents unique health-care

FIGURE 6-6 Gold nanoshells are made to either absorb or scatter light preferentially by varying the size of the particle relative to the wavelength of the light at their optical resonance. These nanoshells can be used to treat cancer in the upper regions of the lung. The nanoparticles specifically target the tumor and, upon excitation with near-infrared light, destroy only the cancerous tissue. SOURCE: Alliance for Nanohealth.

problems, new approaches for early detection and treatment of diseases will be required. Nanomedicine utilizing light-guided and light-activated therapy, with the ability to monitor real-time drug action, will lead to new approaches for more effective and personalized molecular-based therapy.

Nano-sized Sensors and Lighting

Nanospheres also have especially sensitive light-scattering properties that make them ideal for use in chemical sensors. The chemical properties of a tiny object like a biomolecule or a nanosphere can be revealed through Raman scattering of light. Owing to the minute amounts of energy left behind in the molecule, the Raman effect causes a tiny reduction in the frequency of light scattered by a chemical. These Raman frequency shifts are optical fingerprints of the chemical species. Nanospheres can be designed so that chemicals embedded in them or attached to their surface exhibit enhanced Raman scattering by a factor of up to a million. This capability creates the opportunity for many new, all-optical remote-sensing applications sensitive to just a few molecules of a drug molecule, disease protein, or other chemical agent.

The optical properties of nanostructures are also expected to have a significant impact on future energy needs—for example, energy consumption for lighting could be reduced by replacing current technologies (incandescent and fluorescent lighting) with solid-state lighting or OLEDs. The size dependence of optical properties at the nanoscale can also be exploited. For example, while bulk silicon has poor optical properties for optoelectronic devices such as solar cells or lasers, at small dimensions the band gap in silicon can be blue-shifted from the infrared to the visible spectrum; as a result, porous silicon exhibits remarkable room-temperature luminescence. Nanotechnology also has considerable potential to reduce the cost of photovoltaic cells, and recent work has demonstrated that thin-film cells based on nanoparticulate materials can be fabricated with reasonable efficiencies and stabilities.

In the next 10 years, exploration of the nanoworld will likely bring a veritable explosion of new science and technology. AMO physics will play an essential role in this exploration and will also reap the benefits of the developments that come from it.

7

Quantum Information with Light and Atoms

What lies beyond Moore's law? Quantum mechanics and information theory are two of the scientific cornerstones of the 20th century. One describes physics at very small scales, from molecules and atoms to electrons and photons; the other provides a mathematical analysis of data communication and storage. As the last decades have witnessed the remarkable shrinking to near-atomic scales of the electronic components that carry and process information, these two disciplines are naturally beginning to merge. The exponential shrinking of computer chip components, as posited by Moore's law, will soon slow as individual electronic transistors approach the atomic scale, where there is no room for packing more components. However, the revolutionary principles of quantum mechanics could offer a way out: Quantum information science may have profound and far-reaching relevance to economic growth, secure communication, and number-crunching into the 21st century. The quantum hardware now found in atomic, molecular, and optical (AMO) systems is a key to realizing future quantum devices.

THE QUANTUM INFORMATION REVOLUTION

Two of the great scientific, philosophical, and technological revolutions of the 20th century were quantum mechanics and information science. Each of these changed our lives in fundamental and lasting ways. Today we are witnessing the beginning of another revolution, that of quantum information, a revolution that promises similar changes in the 21st century. This new science has come from a merging of the two revolutionary disciplines of the 20th century (see Figure 7-1).

Quantum mechanics describes the world of the very small—the submicroscopic world of elementary particles, electrons, atoms, and molecules, as well as the properties of light at the level of single photons. A revolutionary aspect of quantum theory is its prediction of fundamental ambiguities, where physical properties of objects such as their positions or velocities may coexist in multiple states, a condi-

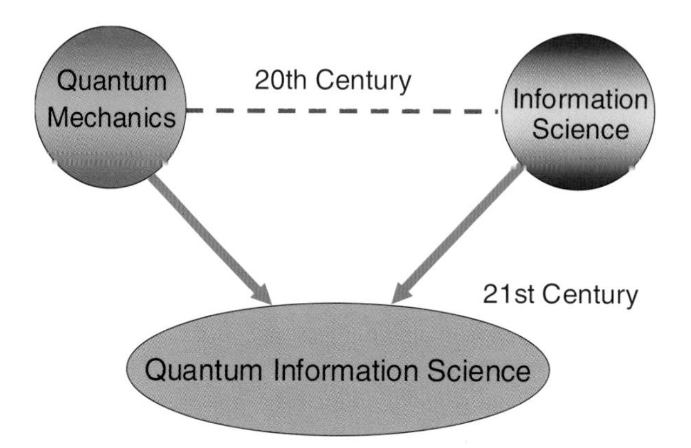

FIGURE 7-1 Two of the profound societal revolutions of the 20th century are combining in the 21st century to create a new science with incredible technological implications.

tion that is inconceivable in our macroscopic world. These ideas are not just arcane academic curiosities, however, but provide the physical basis for chemistry, semiconductor electronics, x rays, and other ubiquitous elements of modern living.

Classical information science describes the storage, transmission, and manipulation of information that is encoded as bits—the ones and zeros of the binary number system. Computers, the Internet, and video games are all products of bit-based information science, and one reason these modern marvels work so well is the nearly complete absence of errors or ambiguities. Bit-based information must be virtually error-free, or else the exponential growth in complexity and speed of computing devices would eventually lead to chaos. Thus, the fundamental ambiguity of quantum mechanics and the fundamental certainty of information science seem totally at odds.

Quantum information changes all of that. A new scientific and technological revolution is emerging in the 21st century out of the new and intimate connection between quantum mechanics and information science. The processing of quantum information requires a physical system that obeys the laws of quantum mechanics. Quantum physics is prevalent in very small, isolated systems such as individual atoms and photons. Thus, AMO physical systems and techniques have taken the major role in the development of quantum information science, just as in the 20th century they were in the vanguard of the development of quantum mechanics. The new quantum information science promises to be as radical in its effect on human society as quantum physics and information science were individually in the last century. In the next 10 years, it will be one of the major driving forces in AMO physics.

WHAT IS INFORMATION?

Information is physical. The storage and processing of information always requires some physical means, such as the orientation of a die, the physical position of a switch, or the amount of electrical charge on a capacitor. In conventional computers information is stored as bits that have two possible (binary) values. Bits are everywhere. Binary information is present in our homes, offices, and cars, contained in literally hundreds of information processors secreted in everyday appliances, in addition to laptop and desktop computers. Our telephone conversations, the music and video entertainment we enjoy, our bank transactions—all involve the storage, transmission, and manipulation of digital information in the form of zeros and ones, represented by billions upon billions of bits. Information is also fungible. These bits take many physical forms, from tiny charges on a transistor, to micron-sized patches of magnetic material, to microscopic burn marks on a CD or DVD. However, all conventional physical bits share one defining feature: A bit is in one state or the other—that is, it is always *either* zero or one but never both.

Quantum information is completely different. Quantum information is stored not in bits but in "qubits," quantum bits whose value can be one or zero but can also be both zero and one at the same time. An ordinary transistor cannot be both on and off! But if it is small enough, so that the rules of quantum mechanics take over, such an oddity is not only possible but is also typical. Thus, a single atom can be in what is known as a "superposition" of two different states. For example, an atom's outermost electron can be spinning with its axis pointing up or down, or it can be in a superposition of up and down. Figure 7-2 shows how this can be represented by a vector having arbitrary orientation in space, in contrast to only two possible orientations for a classical bit. An essential and nonintuitive (!) characteristic of this superposition is that in this state the spin axis is not someplace between up and down. If we observe the spin, it will always be seen to be either up or down (although we cannot predict which it will be beforehand), never someplace in between. But this is not because the spin is indeed up or down before being measured—it is truly in both states and does not appear in one or the other state until it is measured. This very fundamental aspect of the unknown nature of the quantum state until a measurement is made flies in the face of all classical experience and lies at the core of quantum mechanics as we know it today.

Why Quantum Information?

What makes quantum information so special? Superposition allows qubits to do things that ordinary bits cannot. To see why this is so, consider a register of three classical bits (see Figure 7-3). The three-bit register stores the binary number

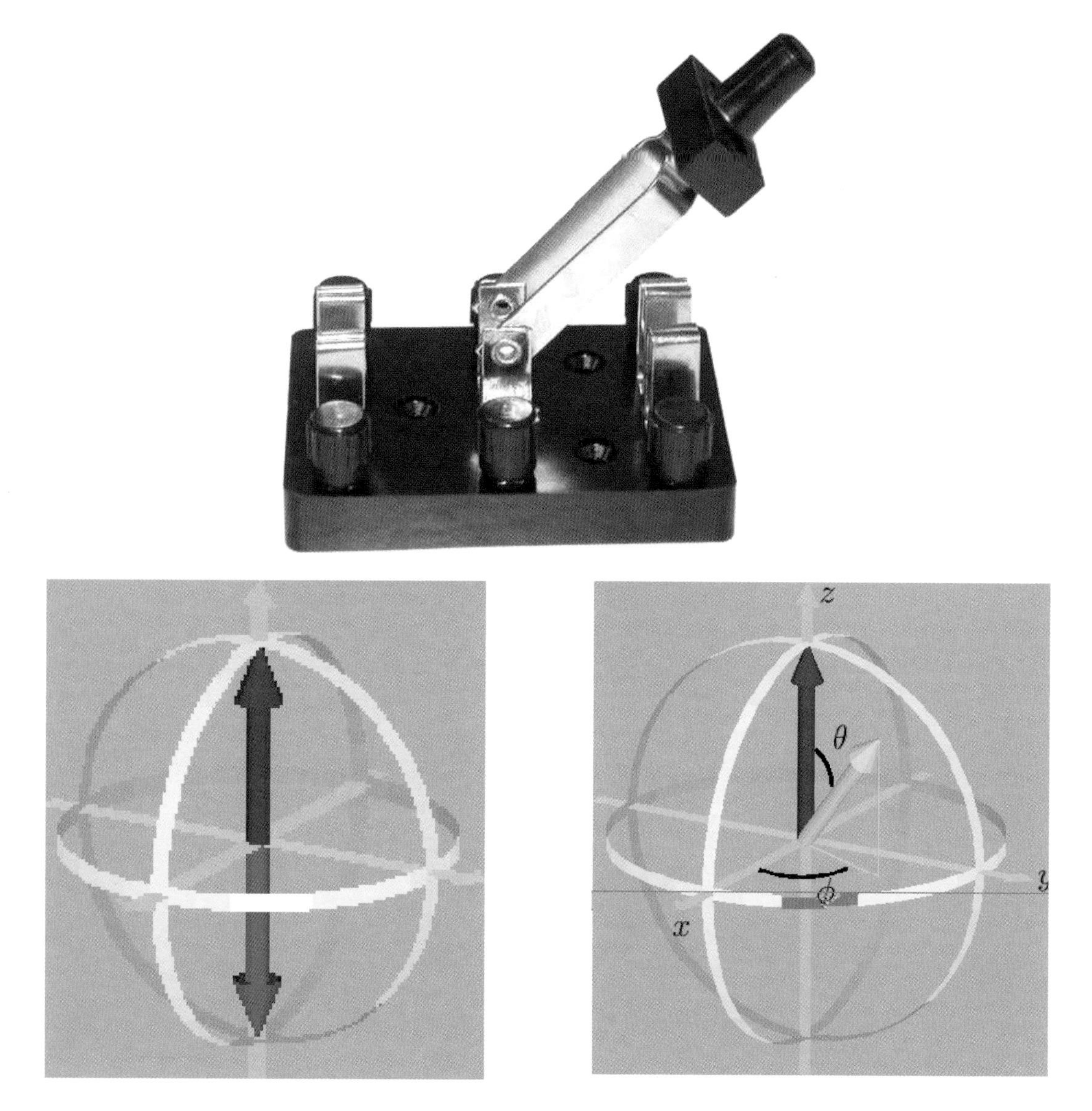

FIGURE 7-2 *Top:* A mechanical switch representing a bit of classical information is either on or off, representing a value of one or zero. SOURCE: L.P. Kelvin. *Bottom left:* A bit can be represented by either up (red arrow) or down (blue arrow) orientations of a classical spin. *Bottom right:* A quantum bit, or "qubit," can be represented by both spin up and spin down at the same time—that is, it can exist in superposition, which is represented by the yellow vector oriented at some direction other than the up and down directions. An electron in an atom can thus provide a quantum switch whose value is indeterminate until measured. SOURCE: R. Laflamme Institute for Quantum Computing, University of Waterloo, Canada.

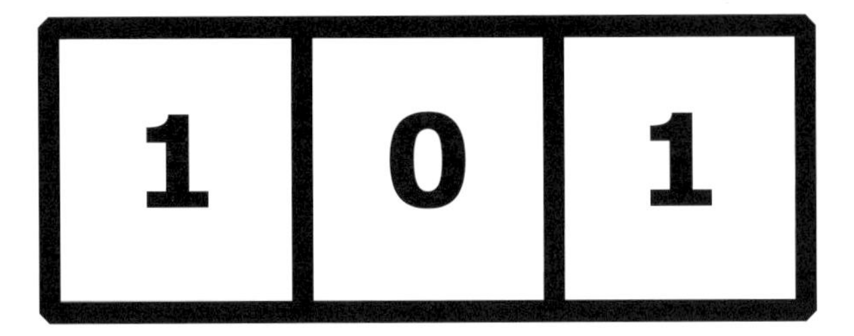

FIGURE 7-3 Three classical bits of binary values 101.

101, which is 5 in the usual decimal system. A three-bit classical register can store any one of eight different numbers from zero = 000 to seven = 111. In contrast, a quantum three-qubit register can store a superposition of all of the eight different numbers at the same time. More precisely, that three-qubit register can be in a coherent superposition of all eight numbers, which we write as follows:

$$|\Psi_{reg}\rangle = a|000\rangle + b|100\rangle + c|010\rangle + d|001\rangle + e|110\rangle + f|101\rangle + g|011\rangle + h|111\rangle$$

This is a shorthand notation for saying that the quantum state of the qubit register $|\Psi_{reg}\rangle$ is a superposition of each of the possible classical states. The variables a through h are related to the relative weights of each state in the superposition. These numbers can take any value, positive, negative, or complex, as long as the squares of their absolute values add up to one. The above superposition is "coherent," because its weights have definite phase relationships between them. This allows interference to occur, much like the interference of any wavelike phenomena (Box 5-2).

The flexibility gained by allowing the qubit register to be in such a quantum superposition is enormous when the number of individual qubits becomes large. The number of states allowed in the superposition grows exponentially with the number of qubits in the register. For the three-qubit register shown, the number of possible states in the superposition is $2^3 = 8$. For an N-qubit register, the number of states in superposition is 2^N. Thus, while a classical N-bit register can represent a single N-bit number, a quantum N-qubit register can be in a superposition of all 2^N N-bit numbers. For a modest 300-qubit register, the number of states in the superposition can be 2^{300}, a number that is enormously larger than the number of atoms in the entire universe.

In addition to this ability to store exponentially many quantum states, the linear nature of quantum mechanics means that these states can all be manipulated at the same time—that is, massive "quantum parallelism" is possible. This is one of the keys to the power of quantum computation. It allows a quantum processor to perform an exponentially large number of calculations all at the same time, since the quantum register contains that large number of different classical registers. This can make a quantum computer mind-bogglingly faster than even the fastest imaginable classical computer. A problem that could take more than a lifetime to solve on the best classical computer of the future might be solved in minutes or hours on a quantum computer. The best example of this today is the problem of factoring a large number into its primes, a computational problem that is extremely time-consuming on a classical computer. In fact, the factorization of large numbers is so hard that it forms the basis of most data encryption standards. In 1994, Peter Shor showed that a quantum computer would be capable of doing this task exponentially faster than any known classical algorithm, i.e., the solution time using the classical algorithm grows exponentially with the number of qubits required to represent the number, while the corresponding time for a quantum solution grows much more slowly. Consequently, quantum computers would compromise the security of many forms of encryption in use today.

QUANTUM INFORMATION AT THE FRONTIERS OF SCIENCE

Quantum information encompasses one of the grand challenges in science in the last century: the reconciliation between quantum physics and classical physics. Here we describe some aspects of this challenge and how quantum information technology from AMO physics may someday bridge this gap.

Despite the dramatic success of quantum mechanics, glaring difficulties remain in reconciling quantum rules of nature with our everyday notions of reality. If quantum mechanics is indeed a complete theory of nature, why does it appear to conflict with classical descriptions of everyday life? Richard Feynman, one of the iconic figures of 20th century physics, memorably stated that

> we have always had a great deal of difficulty in understanding the world view that quantum mechanics represents. Okay, I still get nervous with it. It has not yet become obvious to me that there is no real problem. I cannot define the real problem, therefore I suspect there's no real problem, but I'm not sure there's no real problem.

This "problem" concerns the interpretation of quantum measurements. The superposition principle, telling us that quantum systems can exist in two or more states simultaneously, is by itself not so foreign. After all, any wave phenomenon, such as a sound or a water wave, can exist in many places at the same time and also admits superpositions. But quantum theory goes farther by claiming that the superposition principle for quantum states is only valid in the absence of measurements. When a measurement is made, the superposition randomly "collapses" into one of the definite states that make up the superposition, with a probability given by the weighting of the state measured. For instance, in the three-qubit superposition $|\Psi_{reg}\rangle$ above, the probability of measuring the definite state $|011\rangle$ (decimal 4) is given by the squared magnitude of its weight, $|g|^2$.

The mechanics of this wavefunction collapse, and the resort to probabilities in the measurement of quantum superpositions is perhaps the most revolutionary and bizarre feature of quantum mechanics. The measurement of a quantum superposition, completely different from the behavior of classical wave superpositions, highlights the special role of the observer in quantum phenomena. It recalls the question, Does a falling tree make a sound if no one is there to hear it? Most of us are more comfortable with a physical principle that is independent of whether one (or anything) observes it. Quantum mechanics is the only physical theory that has a special place for the observer, and it is the only physical theory that calls into question our definition of physical reality. Indeed, one interpretation stipulates that whenever a quantum measurement is performed, the universe divides into several universes, each experiencing its own measurement result.

Perhaps the most popular (and usable) melding of quantum mechanics and quantum measurement is the theory of decoherence. This describes how a coherent quantum superposition of states loses its ability to interfere and evolves instead to a classical mixture. In its most common form, decoherence theory applies the usual quantum mechanics to an isolated system and additionally stipulates that when a measurement occurs, the isolation ends and the measuring device (or even the observer) becomes entangled with the original quantum system (see Boxes 7-1 and 7-2). Entanglement with a large object like a measuring device (or a person) enormously increases the complexity of the quantum physics. This increase in complexity causes the nonclassical aspects of the quantum evolution to dissipate, or average out, extremely quickly—far too fast to ever be recaptured. Just as the fragrance that leaves a flower and escapes into the air can never be gathered in once it has dissipated, the quantum aspects of the system appear to become classical once they are measured by a large object. This is the essence of quantum "collapse." Analysis of decoherence allows a practical reconciliation of quantum mechanics with classical physics.

A large-scale quantum information processor can be considered a complex quantum entangled state with many degrees of freedom, but with special architecture or controls that delay the onset of decoherence (see Box 7-2). While such a system could certainly be used for computational applications such as factoring numbers (see Box 7-3) and complex simulations of physical, chemical, and biological systems, the mere survival of coherence at such a large scale may also provide deep insight into the fundamentals of the quantum mechanical description of nature. Can entangled superpositions persist beyond a threshold number of qubits in a real physical system? And apart from the formidable technical prob-

BOX 7-1
Entanglement

The exponential scaling of quantum information is related to one of the weirdest aspects of quantum theory, entanglement (see Figure 7-1-1). This feature is so strange that in 1935, when quantum mechanics was in its infancy, Einstein and two colleagues wrote a paper pointing out just how strange it is. They argued that nothing so strange could be true and instead there must be something wrong with quantum theory. In fact, this was a rare instance where Einstein was wrong—nature is indeed that weird, and we hope to use that weirdness to make useful quantum computers.

Consider just two qubits, each of which can be in one of two states. We can think of these as two atoms, labelled *a* and *b*, whose spins could be either up (state $|\uparrow\rangle$) or down (state $|\downarrow\rangle$) along some chosen direction. Among the possible states of the pair of atoms are these:

$$|\Psi_{sep}\rangle = [|\uparrow_a\rangle + |\downarrow_a\rangle]\ [|\uparrow_b\rangle + |\downarrow_b\rangle]$$
$$|\Psi_{ent}\rangle = |\uparrow_a\rangle|\uparrow_b\rangle + |\downarrow_a\rangle|\downarrow_b\rangle$$

The first of these states is said to be "separable" because it can be written as the product of the state of one atom and the state of the other. In this state, each atom is in a quantum superposition of being up and being down. If we measure the spin of one of the atoms, it will be either up or down (recall that we cannot predict which result will actually be found). We might measure atom *a* to be up and then also find *b* to be down. Or we might find that *a* is up and *b* is also up. The result of each measurement will be random and the result for one atom will not depend on the result for the other. This is a characteristic of separable states. In contrast, the second state cannot be written as the product of the state of one atom and the state of the other. This is an "entangled" state. If we measure the state of atom *a* (or of *b*), it may be up or down, and the result is random. But if we measure *a* to be up, then for this state *b* is bound with certainty to be measured as up. Conversely, if *a* is down, *b* will definitely be found to be down. Thus while the fate of each atom is random, these fates are inextricably linked or correlated. This is what we mean by entanglement.

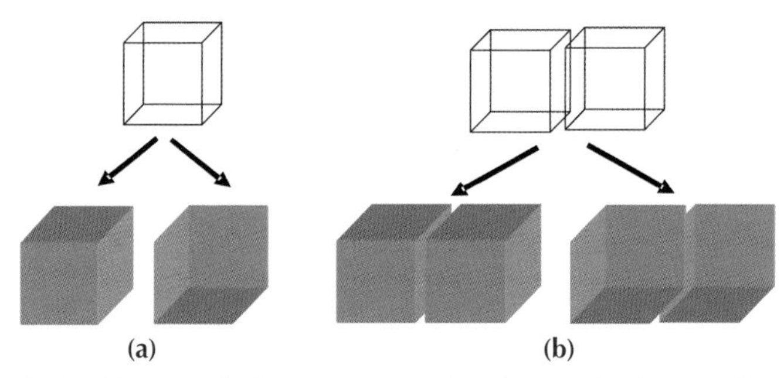

FIGURE 7-1-1 The "ambiguous cube" representations above give a visual sense of superposition, entanglement, and quantum measurement. In (a), the ambiguous perspective of the upper (mesh) cube represents a superposition of a single qubit, and a measurement is analogous to the mind randomly locking onto one of the definite perspectives drawn below. In (b), the ambiguous perspective of two cubes is analogous to an entangled superposition of two qubits. Here, a measurement of either qubit is again akin to locking onto either definite perspective. In this case, however, the other qubit usually appears in the same perspective, even though there is no physical connection between the cubes. SOURCE: Based on figures and descriptions taken from pp. 131, 179, and 180, F.A. Wolf, *Taking the Quantum Leap: The New Physics for Nonscientists*, New York: Harper & Row (1989).

BOX 7-2
Entanglement, Coherence, Schrödinger's Cat, and Decoherence

The most general state of N qubits is an entangled superposition of all 2^N binary numbers: $\gamma_1|000...0\rangle + \gamma_2|000...1\rangle + \cdots + \gamma_{2^N}|111...1\rangle$, with 2^N weights γ_i. For macroscopic or mesoscopic systems, with N larger than 100, questions of formation and control of such entangled states raise many questions about the physics of interacting quantum systems. A very interesting class of such entangled states is constituted by the equal superposition of the two extreme possibilities in which all qubits are either zero or one:

$$|\Psi\rangle_{cat} = |0_1 0_2 0_3 \ldots 0_N\rangle + |1_1 1_2 1_3 \ldots 1_N\rangle$$

The state is referred to as a "Schrödinger cat" state because the two constituents of the superposition are as far apart as possible, analogous to the dead and alive states of the famous feline discussed by Schrödinger in his 1935 analysis of conceptual problems in quantum mechanics:

$$|\Psi\rangle_{cat} \approx |\quad\rangle + |\quad\rangle$$

As Schrödinger himself noted, describing macroscopic concepts such as alive or dead with quantum terms is not necessarily justifiable or useful. However, for mesoscopic systems where N is not too large—for example, between 10 and 1,000—these cat states are useful in both visualizing and characterizing large-scale entanglement. For example, a Schrödinger cat state highlights the difficulty of maintaining complex entangled quantum superpositions, because if just one of the N qubits gets measured by the environment, every qubit will lose its coherence. Any source of decoherence thus makes it more difficult to engineer large entangled states. In fact the survival probability of such states declines exponentially with the number of qubits, posing severe challenges to physical implementation of quantum information processing with large numbers of qubits.

lems that inhibit the construction of a large-scale quantum computer, are there fundamental reasons why such a device cannot be built? Quantum information science is directed at answering these questions. Among the many possible outcomes in this quest, two are earth-shattering. One is that the technical difficulties will be overcome and a useful quantum computer will be built. The second is that in our quest to build such a device, we may discover that the theory of quantum mechanics is incomplete.

QUANTUM INFORMATION TECHNOLOGY

In 1965, Intel cofounder Gordon Moore predicted that the number of electronic components on a computer chip would double every year or two. Moore's law has been remarkably accurate even to this day, when the latest silicon processors now host some 10 million transistors (see Figure 7-4). This exponential growth in

BOX 7-3
Serendipity: Better Atomic Clocks and an Early Quantum Computer

In the early 1990s, experimental AMO physicists in Boulder, Colorado, were investigating the quantum linking of individual trapped atomic ions in order to enhance the signal from atomic clocks based on very few atoms. They determined ways to create certain correlated atomic states that made the atomic clock effectively run faster (see Figure 2-6), resulting in a significantly higher precision. This work was performed at the Time and Frequency Division of NIST, also funded in large part by the Office of Naval Research. These states were precisely the entangled states that are now of central interest to quantum information processing, although that was not realized at the time.

At the 1994 International Conference on Atomic Physics in Boulder, a lecture by an information theorist brought the topic of quantum computation and mathematician Peter Shor's new factoring algorithm to the atomic physics community. AMO theorists at that meeting realized that quantum computers capable of making arbitrary ("universal") computations such as Shor's algorithm could be built with this very same technology of trapped atomic ions. This realization fed into renewed progress on the entangled atomic clock experiments at NIST, so that by early 1995 the experimentalists succeeded in demonstrating the basic step in correlating individual atomic ions in their quest for better atomic clocks by actually implementing the scheme for quantum computation outlined by AMO theorists in the previous year.

The course of this quick turn of events is a case study on how interdisciplinary science happens, how seemingly unrelated areas of research become intertwined, and how funding of fundamental research can lead to very unexpected outcomes. A critical part of these developments involved the interplay between theorists and experimentalists, and between mathematicians, information theorists, and atomic physicists. Quantum information science is now a vigorous activity involving people from nearly all branches of science, mathematics, and engineering.

chip density has been the driving force behind the modern electronic age and has played an instrumental economic role over the last several decades. Will Moore's law continue indefinitely? No. The problem is that adding components by simply expanding the size of chips will slow their speed, so that the only way to sustain this kind of growth is to make transistors ever smaller (assuming that the heat can be dissipated). By the year 2010, Moore's law predicts that a 1 cm^2 chip will have perhaps 10 billion transistors, with each transistor not much larger than a single molecule. At this point, radical change is inevitable. Feynman himself saw far into the future in 1957, evidenced by his famous lecture "There's Plenty of Room at the Bottom," from which the following extract is taken:

> When we get to the very, very small world—say circuits of seven atoms—we have a lot of new things that would happen that represent completely new opportunities for design. Atoms on a small scale behave like nothing on a large scale, for they satisfy the laws of quantum mechanics.

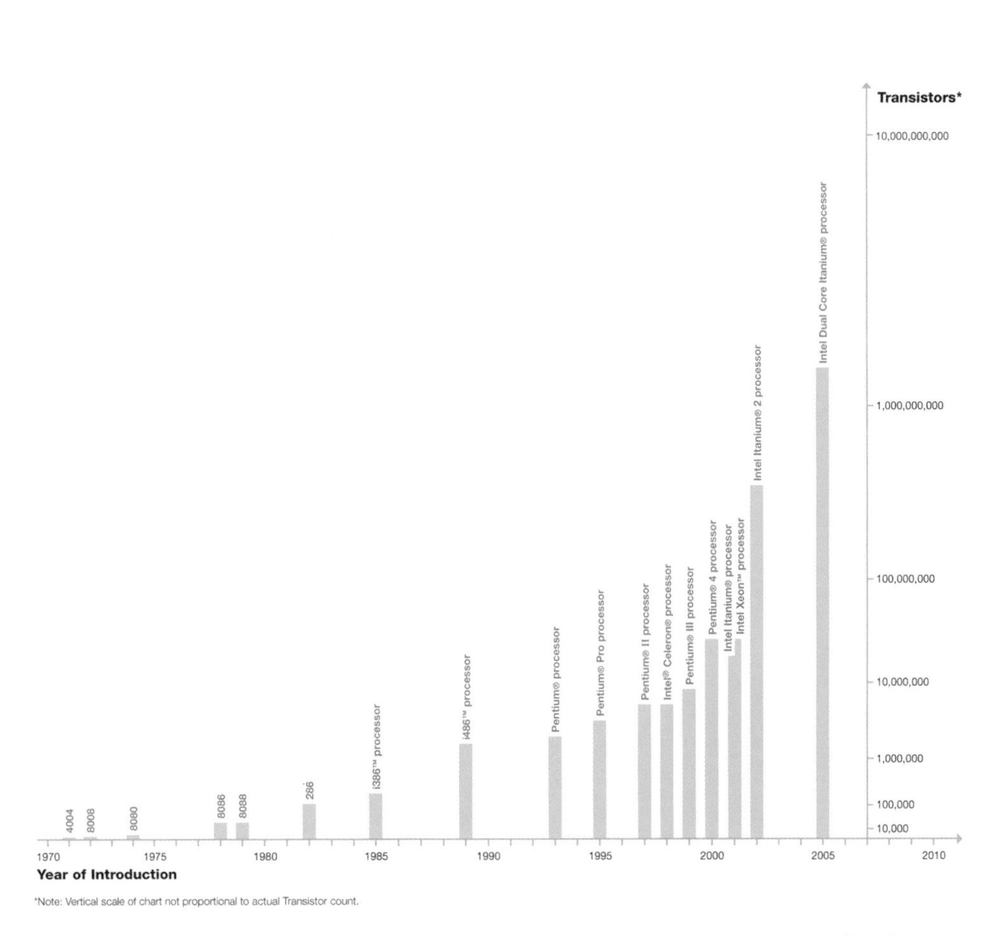

FIGURE 7-4 Processing power, measured in millions of instructions per second, has risen because of increased transistor counts. SOURCE: Copyright © 2005 Intel Corporation.

Quantum mechanics may hold a key to sustained growth in computing power in the coming decades, not by continuing along the path of Moore's law of miniaturization but by changing the rules of the game. This was recognized by several mathematicians, computer scientists, and physicists in the early 1980s and led to the formulation of rules for a computing device operating with qubits that would be manipulated according to the laws of quantum mechanics. Quantum circuits can be drawn up for such a quantum computer. The notion of performing computations quantum mechanically received a major boost in 1994 with the discovery of how to construct a quantum algorithm for the strategic factorization problem (Shor's algorithm, discovered at Bell Labs). The exponential increase in speed that this algorithm shows over the best known classical algorithm circumvents the limitations imposed by Moore's law in a very dramatic manner. For such quantum algorithms to achieve their potential, it is necessary to build a quantum information processor containing many thousands of qubits.

QUANTUM COMMUNICATION

The foundations of information theory were expressed at Bell Labs in 1954 by Claude Shannon, who quantified the concept of information transmitted over a channel (e.g., a series of photons propagating down an optical fiber). Shannon calculated the information content of the photon source, the physical resources needed to store the information, and how much information could be reliably transmitted through a noisy or imperfect channel. These results provide powerful bounds on the maximum possible rate of data compression and the maximum amount of information that can be faithfully transmitted. These results, all for classical information, have had a profound influence on the development of modern communications and information theory.

What benefits might be gained by communicating information with quantum sources and/or quantum channels? Quantum versions of Shannon's results suggest that greater data compression is possible when information is coded into quantum states instead of classical states, and this has been demonstrated in systems of individual photons and atoms. Error correction, through clever coding of additional transmitted information, has also been extended to quantum states. Networks of quantum channels can be used for improved distributed information processing. However, much is still unknown about the capabilities of quantum information.

Quantum Cryptography: A Real-World Application

Coded messages are almost as old as language itself, and today, encryption is essential not only for military and diplomatic purposes but also for industry and commerce. In general, the sender and receiver of the coded message (Alice and Bob in modern parlance) must share some secret key that allows Alice to encrypt the message and Bob to decrypt it. The only completely secure method of encryption is the one-time pad, a random key that connects the plain text to the encoded text. This key must never be reused and is itself as long as the message. Traditionally, Alice and Bob might meet and agree upon the pad to be used, or a trusted agent could be sent from one to the other to distribute the key. The difficulty and inefficiency of such an operation for the transmission of frequent or voluminous coded information is obvious, so other more efficient, but less secure, methods are generally used.

Most key distribution systems in use today derive their security by requiring an eavesdropper to make very hard (and time-consuming) calculations to obtain the key. A simple example is number factoring: given two numbers a computer can very quickly multiply them together to find the product. However, given only the product, the computer must work much harder to find the original numbers. This is the factoring problem mentioned above. Key distribution systems based on this idea are very practical and efficient, but their security is based on the assumption that the eavesdropper has not made advances in computing that are unknown to the users—an assumption that does not always hold. For example, in the 1970s cryptosystems using the original Digital Encryption Standard were thought to be entirely secure but are now easily cracked with modern computers.

Quantum key distribution (QKD) is a new approach that offers unconditional security: The eavesdropper is assumed not to be limited by current technology but to be bound by the known laws of nature. One of the properties of quantum mechanics is that the act of observing, or measuring, a quantum state can cause changes in that state. QKD exploits this property by using quantum systems for key distribution; with the proper choice of quantum states, it is possible to create the situation where an eavesdropper cannot avoid causing changes that can be detected. In this way it is possible to create a natural "wax seal" on the key; as long as no measurment is made, the seal is intact and the key is known to be secure. Figure 7-5 illustrates one method for producing a quantum key.

QKD is most commonly realized with pulses of light containing single photons (the smallest possible amount of light). Single optical-frequency photons on demand are not so easy to make, transport, or detect, but progress toward this goal in the past decade has been very promising. Single-photon QKD systems have been demonstrated over free-space links longer than 20 km, and optical-fiber-based QKD systems now support fairly robust key distribution (10 kilobits per second) over lengths on the order of 10 km. This is already useful for metropolitan-area networks. This may seem slow—it's about the speed of an old telephone modem circa 1980. Indeed, QKD is not likely to ever be as convenient and efficient as conventional key distribution systems. But QKD offers what conventional systems cannot: unconditionally secure key generation. One need not be concerned that advances in decryption technology will reveal one's encrypted messages. This level of security is sufficiently appealing that there are already two commercial ventures selling off-the-shelf QKD systems. The best free-space QKD systems (now only in research labs) can currently support the encryption of streaming video and, as improvements are made in single-photon generation and detection technology, both the bandwidth and reach of QKD systems will increase. There is enough potential for improvement that researchers envision global-scale QKD that supports high-bandwidth, quantum-encrypted networks.

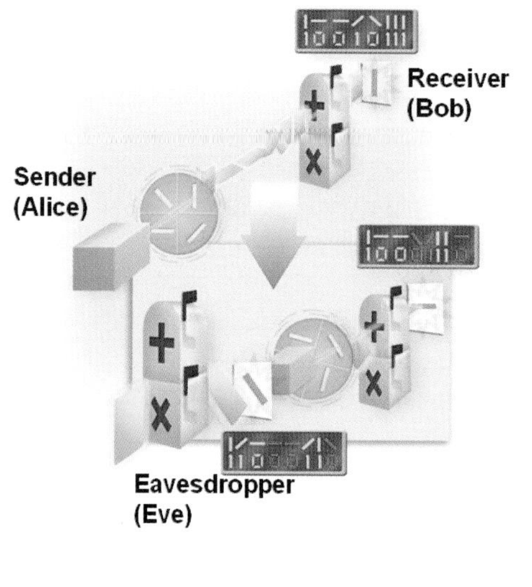

FIGURE 7-5 *Left:* To generate the key, the sender (Alice) randomly chooses one of four polarization orientations for a photon (vertical, horizontal, +45 degrees, or –45 degrees). The photon is sent to the receiver (Bob), who randomly chooses one of two modes (or mailboxes) in which to measure the orientation (vertical/horizontal or +/–45 degrees). After the photons are received, Alice and Bob compare modes, only saving data where their modes match and discarding or ignoring the rest. This guarantees that they will only keep results that are measured correctly. (See final key above Bob's mailbox.) If someone (Eve) tries to measure and resend the photons, errors occur in the data that alert Alice and Bob to the eavesdropping. In the inset picture, Eve tries to receive a vertical photon in the wrong mode and measures it incorrectly as a –45 degree photon. Eve then sends a –45 degree photon to Bob. Bob receives the photon in the vertical/horizontal mode and records a horizontal photon. When Alice and Bob compare their sending and receiving modes they will end up with errors in the key (red data) and thus know that Eve has been listening. SOURCE: National Institute of Standards and Technology, copyright L. Barr. *Right:* Researcher at Los Alamos National Laboratory running a free-space QKD experiment over a 10-km distance between Los Alamos and Pajarito Mountain in New Mexico. SOURCE: R. Hughes, Los Alamos National Laboratory.

Quantum Teleportation Demystified

The notion of transferring an object from one location to another by "teleporting" it in a way that causes it to disappear at the first location and to simultaneously reappear at the second location is familiar to all readers of modern science fiction, and to Star Trek aficionados in particular. The mechanism for such a magical transport of objects is explained as the transfer of all information about the object to an equivalent set of components in the remote location and is usually assumed to result in destruction of the original object. Teleportation of macroscopic or live objects remains science fiction today, not least because of the massive amount of information that would be required to be transferred ($\sim 10^{32}$ bits for a minimal description of the state of a human, which would require over 10^8 centuries to be transmitted by optical fibers). However, the addition of quantum mechanics to this scenario, together with recent experimental advances in AMO physics, has indeed allowed the quantum states of photons and atoms to be teleported, for atoms over very short distances and for photons over distances of several kilometers (see Figure 7-6). Photon teleportation over global distances via satellite links is now under consideration.

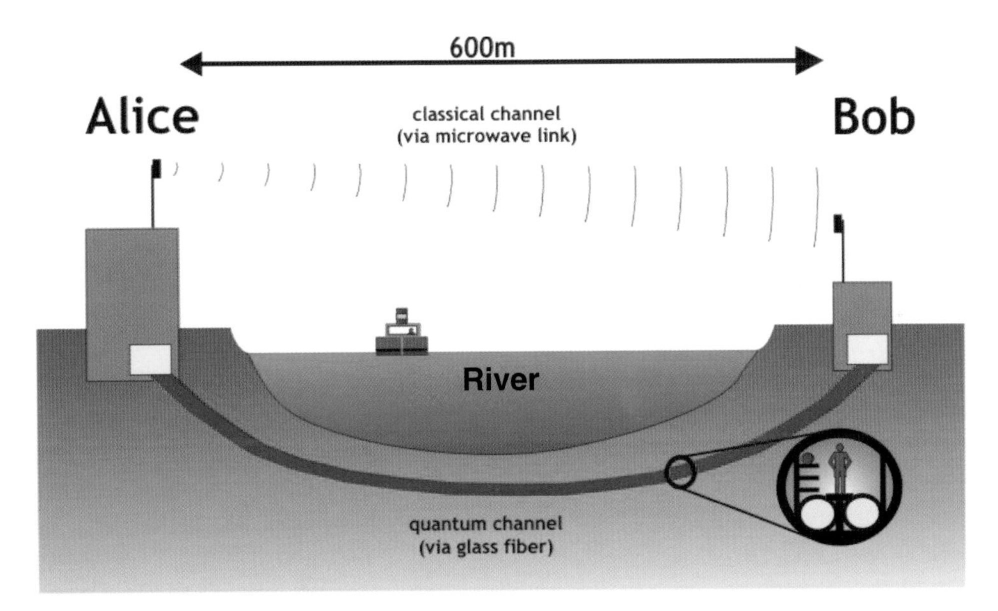

FIGURE 7-6 Quantum teleportation experiment demonstrated using an 800-m-long optical fiber fed through public sewer system tunnels under a river. Quantum teleportation is very different from the transporter described in Star Trek. In a quantum teleportation only the quantum state is transferred and not the qubit itself—that is, the atom or photon constituting the two-level quantum system is not transferred. This is very different from the image of a disappearing Captain Kirk when all of the information contained in his vast collection of constituent atoms is scanned by a mysterious transporter. Furthermore, quantum teleportation cannot occur instantaneously, unlike the commonly assumed science fictional mode of intergalactic personnel teleportation. The speed of quantum teleportation is limited by factors such as the speed at which the necessary entanglement can be generated and measurements made, as well as classical transmission of the result of the two-qubit measurement. Thus there is no violation of the laws of relativity. SOURCE: Copyright © Institut für Experimentalphysik, University of Vienna, Austria.

Quantum teleportation provides a route to perfect transfer of quantum states between spatially separated locations, without transfer of matter. Sometimes referred to as the "disembodied transfer of quantum information from one place to another," it has already found numerous applications in quantum communication and computation. It could become the basis for a quantum internet and might be very useful in overcoming architectural constraints in construction of quantum computers. The origin and destination points need only be connected by a classical communication channel, and they must also share entanglement. This introduces the quantum nonlocality that allows the direct transfer of quantum information without sending it through the classical channel. Methods for producing entangled pairs of photons are becoming a standard tool of quantum optics.

What is the limit to quantum teleportation? All experiments to date have teleported states of single qubits. In principle, multiqubit states could also be teleported, although more complex measurements would be required. "Teleporting" the complete quantum state of a complex molecule with internal dynamics might constitute the next stage, perhaps ultimately approaching even transfer of the quantum information describing the complete state of a small biological molecule. The difficult synchronization requirements this would introduce are as yet unaddressed. In the meantime, the use of quantum teleportation for moving states of atoms and photons around with high fidelity promises significant advantages for any schemes manipulating quantum states for metrology and high-precision measurements, or for facilitating data transfer in quantum information processing.

VISION FOR LARGE-SCALE QUANTUM HARDWARE

Shannon's definition of information in terms of bits led to the experimental representation of bits in nature, from unwieldy vacuum tubes in the mid-20th century to the modern semiconductor transistors of less than 0.1 micrometers in size. In the 21st century, we find the new theory of quantum information is stimulating the development of a very high level quantum control of physical systems, resonating with the goal of modern AMO—to control our quantum world. The main hardware requirements for a quantum information processor are these:

- *Requirement 1.* The quantum system (i.e., a collection of qubits) must be initialized in a well-defined quantum state.
- *Requirement 2.* Sufficient operations must be available and controlled to launch the initial state to an arbitrary entangled state.
- *Requirement 3.* Measurements of the qubits must be performed with high quantum efficiency.

From Requirements 1 and 2, the qubits must be well isolated from the environment to ensure pure initial quantum states and to preserve their superposition character, but they must also interact strongly with one another in order to become entangled. On the other hand, Requirement 3 calls for the strongest possible interaction with the environment to be switched on at will.

These severe hardware requirements rule out most known physical systems. The most advanced candidates for quantum information processors in the next decade will likely come from atomic, molecular, and optical physics. Many solid-state architectures hold promise in the long run, and the quantum control of simple solid-state degrees of freedom is now beginning to take hold, but the environmental isolation afforded by individual atoms and photons is unsurpassed by that of other physical systems. Complex quantum entangled states of AMO systems can be generated by controlled quantum operations on small numbers of atoms with lasers. Such operations can be used to construct quantum circuits, the analog of Boolean circuits for classical computation, with classical gates replaced by corresponding reversible quantum logic gates. The complex global entangled states that can exist in strongly correlated materials such as Bose condensates and Fermi gases might also be useful for quantum computing, given that each atom in the collection can be independently measured, enabling an alternative paradigm for quantum computation.

The following subsections summarize several implementations of qubits and basic elements of future quantum information processors that are composed of individual atoms and photons. Typically, qubit memories will be formed from atoms or molecules, with links between qubits established by electromagnetic interactions between the atomic or molecular memories.

Trapped Atomic Ions

Individual atomic ions can be confined with electromagnetic fields to be nearly motionless, forming well-defined crystals of atoms separated from each other through their mutual Coulomb repulsion. The isolation of trapped ions, along with their strongly coupled motion, makes them a natural candidate for quantum hardware. Trapped ions can be entangled following a number of schemes, most involving laser beams that apply a force to the ion crystal depending on the internal state of one or more of the trapped-ion qubits. Moreover, the qubit states within trapped ions can routinely be measured with near-perfect efficiency.

Nearly all of the rudiments of quantum information processing have already been demonstrated at the few-atom level in ion traps, and the future will see the extension of these systems to hundreds and thousands of atomic qubits. With a growing number of trapped-ion qubits, it will become difficult to control the motion of such a complex crystal as the system approaches a solid-state system with large thermal vibrations in too many degrees of freedom. Instead, one idea for scaling the trapped-ion quantum computer is to entangle only a small number of ions at a time in very simple crystals, while shuttling ions between zones to connect with other ions. Figure 7-7 depicts visions of ion trap architectures that electromagnetically shuttle individual ions between various trap zones in a very complex "quantum computer chip" composed of semiconductor electrodes that are fabricated much like current silicon chips.

Optical Lattices

Neutral atoms can be confined similarly to atomic ions, but because they have no electric charge, they can only be held with laser beams or magnetic fields. In the context of quantum information processing, the optical lattices described

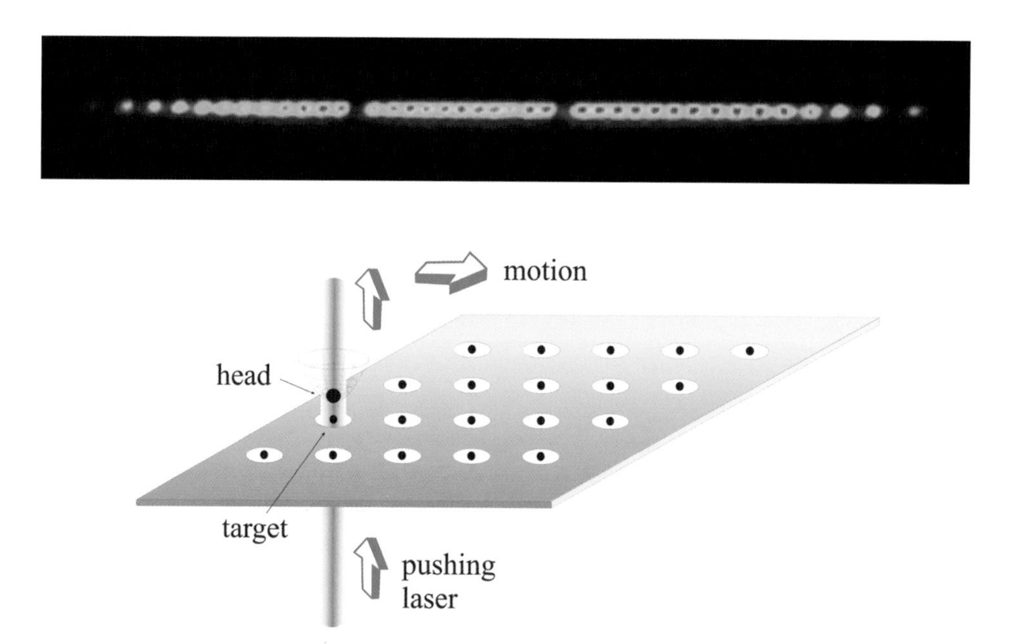

FIGURE 7-7 *Top:* Array of a few dozen trapped mercury ions. The ions fluoresce from applied laser radiation, and the apparent gaps are different isotopes that fluoresce at different wavelengths. The distance between the first and last ions is about 1 mm. SOURCE: J. Bergquist and D. Wineland, National Institute of Standards and Technology. *Bottom:* Array of stationary trapped-ion quantum memories, with a single roving head ion interacting with any memory ion in order to move quantum information between memories. SOURCE: J.I. Cirac and P. Zoller, University of Innsbruck, Austria.

in Chapter 3 present a promising avenue for large-scale applications. The use of quantum degenerate gases (Bose condensates or Fermi gases) may allow loading of the individual lattice sites with exactly one (or two, or three, or more) atoms per site. The separation between atoms in an optical lattice is thousands of times larger than a typical separation of atoms in a solid-state crystal and offers the possibility of doing controlled operations on qubits that are defined in terms of internal electronic states of the atoms. Figure 7-8 shows how atoms trapped in a two-dimensional optical lattice can be manipulated by controlling the laser fields. Knobs representing the laser strengths and phases are turned to change the shape of the lattice, bringing atoms into contact with their neighbors. Turning these knobs in different ways allows interactions with any of the four neighbors. Notice that massive parallelism is designed into this particular manipulation. In certain lattices it is also possible to address and manipulate individual atoms.

Optical lattices present a fascinating possible way to avoid the decoherence that limits the realization of a practical quantum computer, through creation of entangled states of atoms. Atoms produced in topologically protected quantum states—that is, quantum superpositions of states whose relative phase is embodied only in many-particle, nonlocal correlations—cannot have their states randomized by local interactions or interactions involving only a single particle, which is generally how the environment causes decoherence. Such entangled states can therefore be more robust against decoherence.

Solid-State Quantum Bits

There are now several solid-state systems that have successfully reproduced the quantum behavior typical of AMO systems. These have involved confined superconducting systems where either the magnetic flux or the electric charge is quantized (see Figure 7-9). This exciting development underscores the inherent interdisciplinary nature of the field of quantum information science. In particular, as was also discussed in Chapter 3, more and more links between AMO and condensed matter physics are materializing, providing fertile new ground for advances in both areas.

Photonic Qubits

Individual photons are notoriously difficult to produce in a controllable fashion. Even though a simple attenuated light source can be made to produce primarily single photons, it is impossible to know exactly when a photon appears. To produce single photons on demand, the photon is typically produced by some matter quantum system, such as a single atom or a single semiconductor quantum dot.

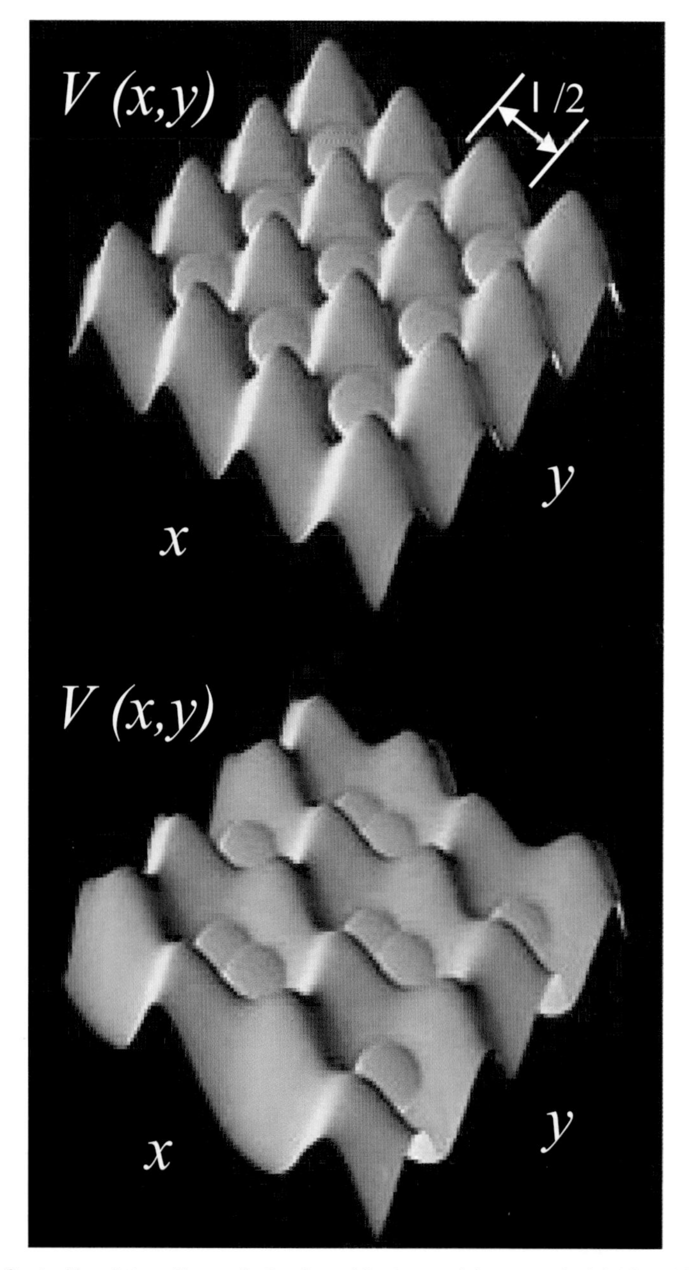

FIGURE 7-8 Controlling interactions of atomic qubits trapped in an optical lattice. *Top:* Individual atoms (purple) are confined in two-dimensional potential wells formed by crisscrossed laser beams, spaced by one half of the optical wavelength. *Bottom:* By controlling the parameters of the external lasers, atoms can be made to approach and interact with their neighbors. SOURCE: National Institute of Standards and Technology.

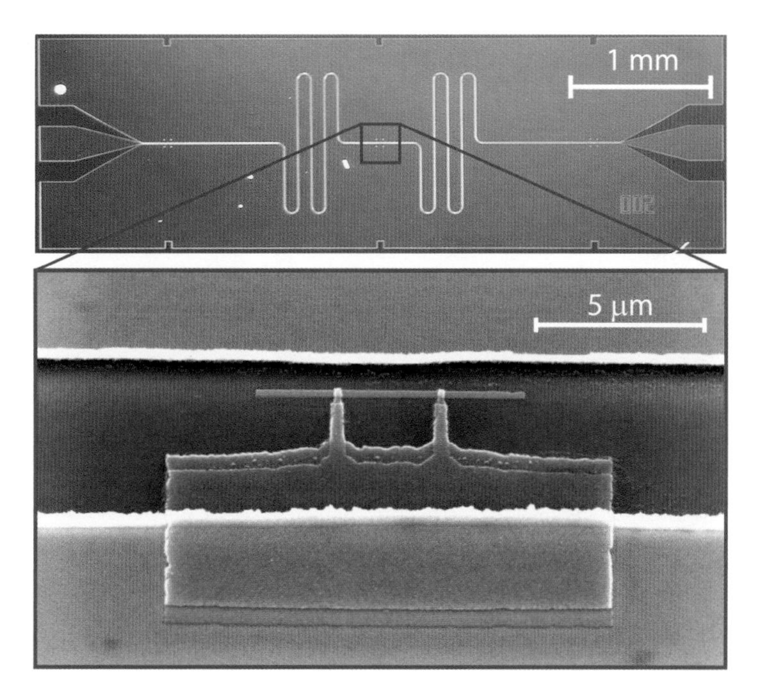

FIGURE 7-9 An example of a superconducting circuit which is used for quantum information processing. A quantum bit made from a Cooper-pair box, consisting of two pieces of superconductor (purple), coupled together by a pair of Josephson junctions (white squares) in a small loop. Manipulation of the quantum states of the qubit are made by transferring single Cooper pairs across the junctions, under the influence of electric fields applied to a gate lead (beige leads on above and below) and magnetic fields through the loop. Such qubits can also be integrated with microwave transmission line resonators (top) to realize a solid-state testbed for quantum optics, called circuit quantum electrodynamics, where single microwave photons can be coherently coupled to qubits. These and other solid-state technologies have great potential for the future of quantum information processing. SOURCE: R. Schoelkopf, Yale University.

Once this is accomplished, these single-photon sources can be used for a variety of applications, ranging from secure quantum cryptography to full-blown quantum computing. One source of entangled photons that has been used for many seminal demonstrations of quantum information applications is the nonlinear downconverter, where a single (blue) photon entering a nonlinear crystal produces twin (red) photons at the output, as depicted in the Figure 7-10.

Qubit Converters Between Atoms and Photons

In classical information processing, electrical current communicates information from place to place, and magnetic domains or electrical voltages act as

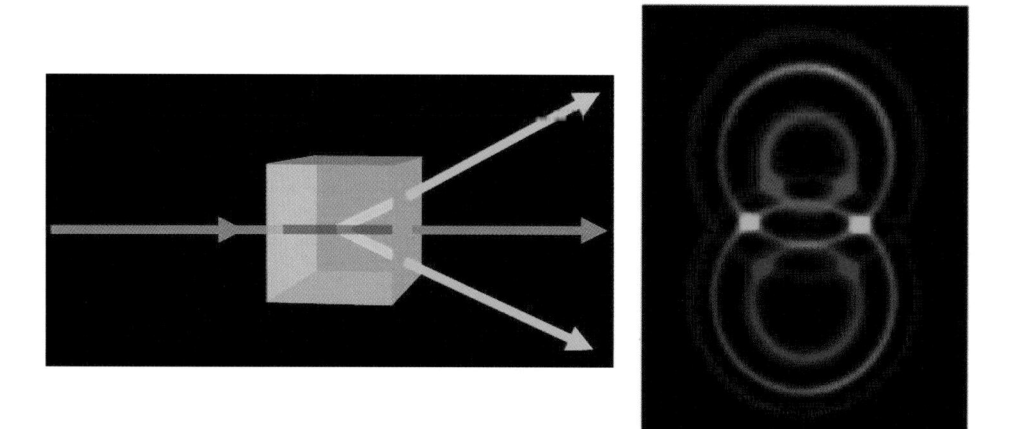

FIGURE 7-10 *Left:* When an ultraviolet laser beam (purple) passes through a nonlinear crystal, the photons in the beam can interact with the crystal to produce two daughter photons in the visible spectrum (each green, for example). *Right (view looking into laser beam):* When a laser beam sequentially traverses two specially oriented crystals, many colors are produced, with each crystal producing a series of concentric circular patterns of light. Considering only the directions where the green circles intersect, it is unknown which of the two green photons came from which crystal, and this can produce entanglement between the two photons. Such sources are the workhorse for optical implementations of quantum information protocols. SOURCE: A. Zeilinger, University of Vienna, Austria, Copyright © Institut für Experimentalphysik, University of Vienna, Austria. Photo: P. Kwiat and M. Reck.

memory elements where the information is stored. The fundamental device that links storage and communication is the transistor, depicted in Figure 7-11, a switch for electrical current that is controlled by the amount of charge stored in a memory. Throughout the second half of the 20th century, the development and refinement of the transistor catalyzed the information era, from the original unwieldy vacuum tubes in the 1940s and the first digital computers to the solid-state transistor developed in the 1950s and ultimately miniaturized to submicrometer (10^{-6} m) dimensions.

In quantum information processing, storage and communication of qubits will likely take far different forms. Good quantum memories should be localized in space and quite isolated from the environment, while good quantum communication resources must travel easily between locations with minimal environmental disturbance. AMO physics presents some nearly ideal candidates for quantum memories (atoms) and ideal quantum communication channels (photons) and is therefore at the forefront of the development of good quantum information hardware. A grand challenge in AMO science is the development of a "quantum transistor" that interconverts quantum memories of individual atoms to quantum

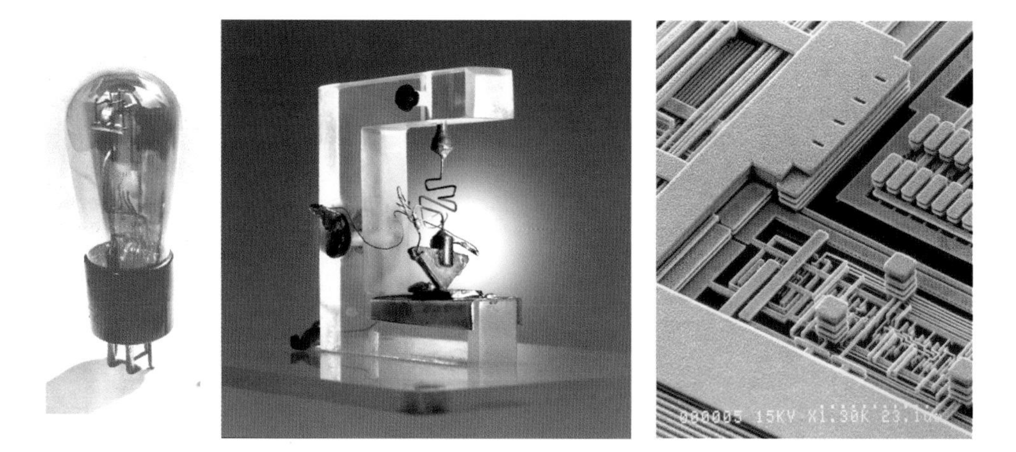

FIGURE 7-11 *Left:* Vacuum-tube transistor. SOURCE: Lucent Technologies. *Center:* First solid-state transistor, developed by Bardeen, Brattain, and Shockley in 1947. SOURCE: Lucent Technologies. *Right:* Modern microscale transisitors on an integrated circuit. SOURCE: IBM.

information channels of individual photons. Simple interconversion methods could lead to the development of quantum networks of qubits and distributed quantum computing, where small numbers of atomic qubit memories are linked within a single device or between remotely located devices with optical fibers. As photonic quantum information propagates through optical fibers over very long distances—say, across a continent or through space—the quantum signal will eventually degrade. However, while conventional (classical) signals can be boosted through amplification, quantum information cannot be amplified without adding significant noise to the system. This is a result of the quantum "no-cloning" theorem—that is, the fundamental impossibility of reproducing an unknown quantum state. A quantum amplifier is tantamount to a measurement that destroys the information itself. Nevertheless, quantum information can indeed be "purified" as it slowly degrades through a long communication channel by converting photonic qubits to atomic qubit memories and implementing small quantum computers at "quantum repeater" nodes that provide quantum error correction (see Figure 7-12).

One natural approach for linking atomic and photonic qubits is to trap single atoms and single photons simultaneously in the same space. Here, the photon is itself confined between two highly reflecting mirrors for long enough so that multiple atoms can interact with the same photon. The technical requirements for such cavity mirrors are extreme, sometimes requiring mirror losses and transmission on

FIGURE 7-12 Optical equipment at the Georgia Institute of Technology is used to transfer information from two different groups of atoms onto a single photon in a prototypical quantum network node or quantum repeater. A quantum repeater converts photons into atomic quantum memories at periodic intervals, whereby small quantum computers can purify the quantum information without destroying it, before it is sent out again. This is similar to the use of conventional repeater circuits, which can, for example, deliver power over large distances. Quantum repeater networks allow stored quantum information to propagate over long distances despite noise and losses en route. Single photons can typically travel no more than about 10 km before they are lost. SOURCE: A. Kuzmich, Georgia Institute of Technology.

the order of one part per million. Nevertheless, there is constant progress in the controlled interaction between atoms and photons within optical cavities.

In these cavities, qubit states stored within the atoms can be interconverted to photon states in the cavity with the use of external laser beams, as depicted in Figure 7-13. The photon can be made to leak out of the cavity within a preset time window, resulting in an ideal source of single photons for use in quantum communication. Moreover, after the photon leaks out of the cavity, it can be caught in a second cavity by the application of an appropriate laser pulse in the second cavity. Generalizations involving more than one atom in each cavity can distribute entanglement to many nodes.

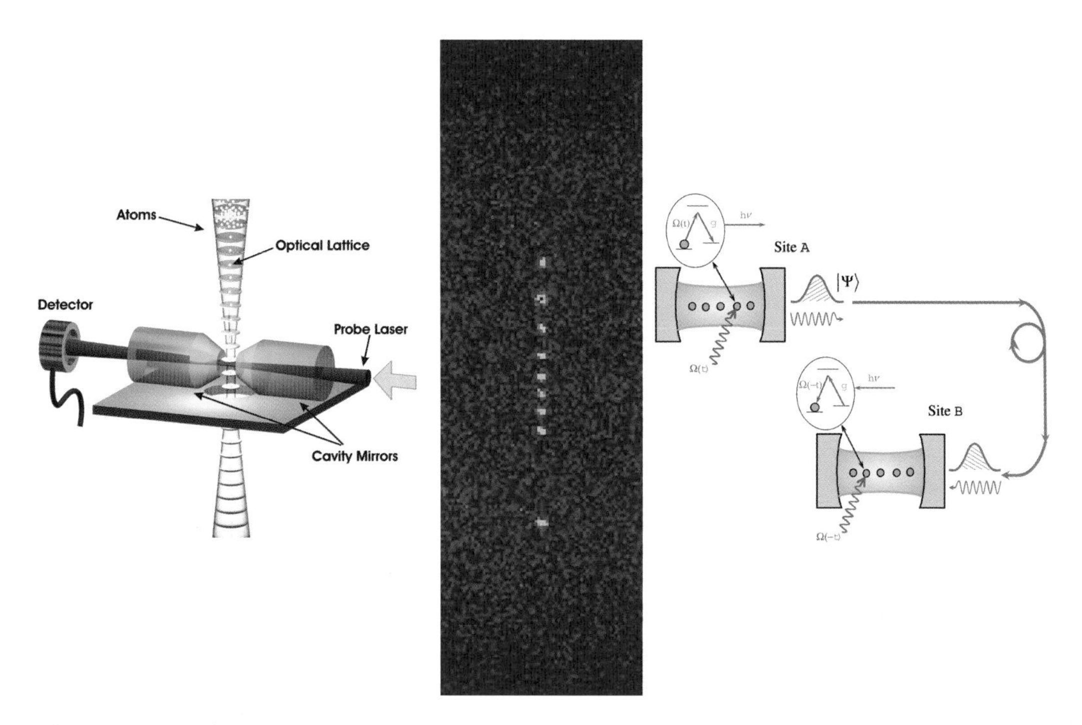

FIGURE 7-13 *Left:* Schematic of atoms confined in a laser beam "conveyor belt" being loaded into an optical cavity. Quantum information from each atom can be transferred to the state of photons in the cavity, which can subsequently be mapped onto other atoms. SOURCE: M. Chapman, Georgia Institute of Technology. *Center:* Image of individual neutral atoms confined in a one-dimensional lattice formed by counterpropagating laser beams. SOURCE: M. Chapman, Georgia Institute of Technology. *Right:* An alternative method for linking atoms through photons is to allow the photons' quantum information to leak controllably through the mirrors and carry the quantum information to another cavity, possibly remote. SOURCE: H.J. Kimble, California Insitute of Technology.

WHAT WOULD WE WANT TO COMPUTE WITH A QUANTUM PROCESSOR?

Quantum computers derive their power through quantum parallelism, as discussed above. Not only can N quantum bits store a superposition of all 2^N binary numbers, but also, when a quantum computation is performed on this superposition, each piece gets processed simultaneously. For example, quantum logic operations can shift all of the bits one position to the left, equivalent to multiplying the input by 2. When the input state is in superposition, all inputs are simultaneously doubled with one turn of the crank (see Figure 7-14).

After this quantum parallel processing, the state of the qubits must ultimately be measured. Herein lies the difficulty in designing useful quantum computing

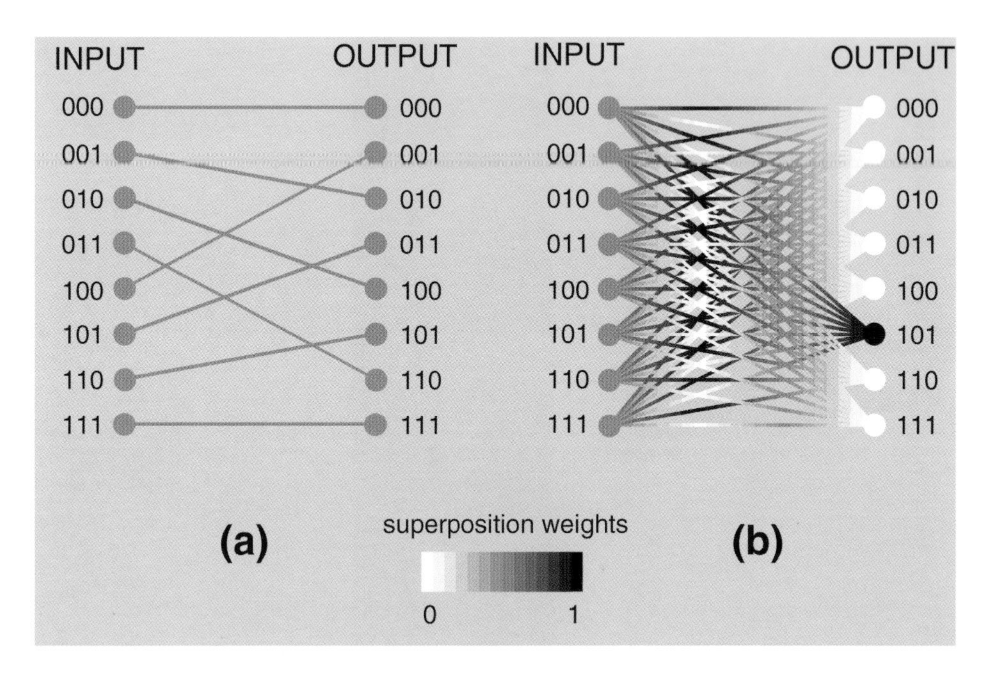

FIGURE 7-14 Simplified evolution during a quantum algorithm on $N = 3$ quantum bits. The inputs are prepared in superposition states of all $2^N = 8$ possible numbers (written in binary). The weights of the superposition are denoted on the gray scale, where black is a 100 percent weight and white is a 0 percent weight. (a) Quantum algorithm for simultaneously doubling all input numbers by shifting all qubits one position to the left and wrapping around the leftmost bit. The outputs are also in superposition, and a final measurement projects one answer at random. (b) Quantum algorithm involving wavelike interference of weights. Here, quantum logic gates cause the input superposition to interfere, ultimately canceling all of the weights except for one (101 in the figure), which can then be measured. For some algorithms, this lone answer (or the distribution of a few answers after repeated runs) can depend on the weights of all 2^N input states, leading to an exponential speedup over classical computers.

algorithms: According to the laws of quantum mechanics, this measurement yields just one answer out of the 2^N possibilities; worse still, there is no way of knowing which answer will appear. It seems that quantum computers offer no advantage in computing functions where each input results in a unique output, as in the doubling algorithm above. The trick behind a useful quantum computer algorithm involves the phenomenon of quantum interference and delaying the measurement. The weights in a coherent quantum superposition are wavelike, so they can be made to interfere with each other. Some weights interfere constructively, like the crests of an ocean wave, while others cancel, as when a valley meets a crest. In the end, the parallel inputs are processed with quantum logic gates so that almost all of the weights cancel, leaving only a very small number of answers, or even a single answer, as depicted in Figure 7-14b. By measuring this answer (or repeating the computation a few times and recording the distribution of answers), information can be gained pertaining to all 2^N inputs. This answer (or sparse distribution of answers) can, in some cases, carry global information pertaining to all inputs.

The best known algorithm that can be cast in this form is Shor's quantum factoring algorithm, described in Box 7-4. Factoring is the reverse of multiplication: Given a number, M, we are interested in finding numbers p and q that produce M when multiplied ($M = p \times q$). For instance,

$$163780975899 \times 89400273645 = 14642064063215749881855$$

But,

$$172668737908098194191 3973 = \underline{?} \times \underline{?}$$

Determining the factors of a number is a difficult problem, and the difficulty scales exponentially with the size of M. Demonstrating this difficulty, it is noteworthy that in 2005, 80 supercomputers took more than a year to factor a 200-digit number. Given the same resources, factoring a 500-digit number would take over a trillion years.

The extreme difficulty in factoring large numbers is precisely the reason that commonly used encryption schemes remain safe. The most popular encryption scheme is RSA public key cryptography, named after its inventors, R. Rivest, A. Shamir, and L. Adleman. Unlike one-time keys, which are intrinsically secret, here the cryptographic key is made public. A very large number—possessing perhaps 300 decimal digits—is broadcast. The factors of this key are known only by the recipient (Bob) of a message. The sender (Alice) encodes her message in a special way using Bob's publicly available key, and because Bob is the only person on Earth who knows the factors of his key, he is the only one who can decrypt Alice's message (see Figure 7-15).

As mentioned above, Shor's quantum algorithm for factoring would factor large numbers exponentially faster than any known classical algorithm.[1] This quantum algorithm has not yet been implemented for anything approaching an interesting number. Indeed, factoring a number with 300 digits would require near-perfect control of at least 10^6 quantum bits and perhaps 10^9 near-perfect

[1]It is important to realize that classical resources (e.g., computer speed and capacity) are themselves improving at an astonishing rate. For instance, a new computer hardware design exists that could factor 300-digit numbers in less than a year, so that factoring a 500-digit number might not be so far off.

BOX 7-4
Shor's Quantum Factoring Algorithm

The breakthrough discovery of the quantum factoring algorithm by Peter Shor in 1994 involves a juxtaposition of the disjoint disciplines of number theory and quantum physics.

There is no known efficient method for factorizing a number N into a product of two smaller numbers, $N = p \times q$. The fastest classical algorithms effectively amount to a brute-force elimination of prime numbers following repeated division tests (is N divisible by 2? By 3? By 5? By 7? and so on), resulting in an exponential scaling of computation time with the size of the number N (as measured in the number of bits).

An alternative, but equally slow, method of factorization is to search for the periodicity of the function

$$f(x) = a^x \pmod{N}$$

where a is an arbitrary constant (sharing no common divisors with N) and the "mod N" notation tells us to take the remainder of a^x after dividing by N. The function $f(x)$ is indeed periodic with x, but it is not at all obvious what the period is. In fact, it has been known for centuries that the period r of $f(x)$ is closely related to the factors p and q of N. This is because by simple substitution of $f(x + r) = f(x)$ into the preceding equation, we find

$$a^r = 1 \pmod{N}$$

or, equivalently, N divides (a^r1) evenly, so the factor p (or q) divides $(a^{r/2} \pm 1)$. Thus, we find that both N and $(a^{r/2} \pm 1)$ share p (or q) as a common divisor, and Euclid's algorithm for finding the greatest common divisor of two numbers can be used to efficiently deduce the factor p.

Unfortunately, finding the period of $f(x) = a^x \pmod{N}$ in the first place is tantamount to performing on the order of N^2 brute-force evaluations of $f(x)$ and checking for when the function repeats, once more requiring a computing time that scales exponentially with the number of bits required to represent the number N.

Finding the period of a function is exactly the type of application that quantum computers can do very efficiently. Quantum computers can evaluate an exponential number of inputs simultaneously through the superposition principle. Of course, if a measurement is performed at this stage, a useless random result will be recorded. But the period of a function is a global property, so that by calculating $f(x)$ for all inputs (up to about N^2) simultaneously, it would seem possible to shuffle things so that this joint property could be extracted. Indeed, by performing a Fourier transform on this periodic function, the spectrum of results is only sparsely populated, meaning that nearly all of the weights of the resulting quantum superposition are almost zero. Now, following a measurement of the system (actually a distribution of a relatively small number of measurements), the period can be determined with high probability. In the end, the required time (for both quantum and ancillary classical computations in the algorithm) scales only as a polynomial of the number of bits required to represent N. This represents an exponential reduction in the time compared with the time for the execution of a classical factoring algorithm.

quantum logic gates, the analog of classical Boolean operations. In addition, if some overhead is included to ensure error correction during the factoring process, the factoring computer is estimated to require a whopping 10^9 quantum bits and 10^{14} quantum gates. These numbers are dauntingly large and will not likely be reached

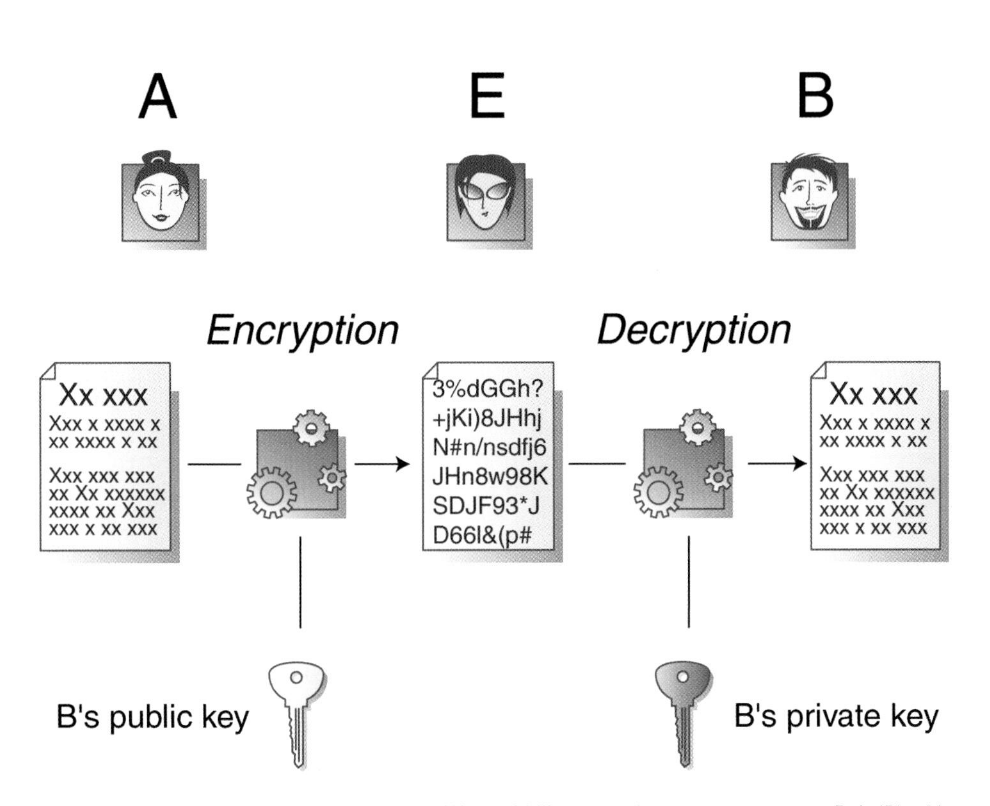

FIGURE 7-15 Public key cryptography. Alice (A) would like to send a secret message to Bob (B) without Eve (E) being able to read the message. Alice encrypts a document using Bob's public key. The encrypted document is unreadable to Eve and to anyone else but Bob, who can decrypt the document using his private knowledge of the prime factors of the public key. SOURCE: Cryptomathic Ltd.

in the coming decades. However, the mere existence of Shor's algorithm demands a long-term perspective on the future evolution of information processing. The possibility of having a full-blown quantum computer in 50 years should impact today's encryption standards, especially when applied to information that must be kept secure for long periods of time.

Another quantum algorithm that has received considerable attention is the quantum algorithm for searching a database, Grover's algorithm, named after its inventor, also from Bell Laboratories. Suppose you want to find the person in an alphabetized phone book who has a given phone number. Because the phone numbers are completely uncorrelated with the spelling of the name, you would have to look through half of the phone book on average—$N/2$—before finding the correct name. Grover's algorithm uses quantum mechanics to succeed in the search after examining only $\sqrt{N}$ items in the list, a quadratic speeding up. For example, using a classical computer to search for a single item in a database containing a million items would typically require 500,000 steps. This effort would require only 1,000 steps with the quantum search. This speeding up offered by a quantum search is again due to quantum parallelism, which permits us to search the entire database at once. Grover's search algorithm has been implemented in very small quantum computers built from quantum components spanning all realms of AMO physics: electrons within individual atoms, trapped atomic ions, complex molecules, nuclear spins, and individual photons. Although these are all rudimentary implementations of quantum search, they exemplify the flexibility of AMO systems for quantum information applications and also indicate how to scale up to more complex versions of this and other quantum algorithms in any physical system.

The quantum search algorithm has broad applicability to many problems in computer science and our information-oriented society beyond searching of large databases. For example, Grover's algorithm can be adapted to the task of quickly finding two equal items among *N* given items. More generally, the quantum search algorithm can offer a significant speeding up of any application involving exhaustive trial-and-error or brute-force solution techniques. This may have the most notable impact in exhaustive searches for the solution to so-called "NP" problems, whose solutions can be efficiently verified but not easily found and which have long confounded computer scientists. To quote mathematician and Fields medalist Michael Freedman of Microsoft Corporation,

> Setting aside the constraints of any particular computational model, the creation of a physical device capable of brutally solving NP problems would have the broadest consequences. Among its minor applications it would supersede intelligent, even artificially intelligent, proof finding with an omniscience not possessing or needing understanding. Whether such a device is possible or even in principle consistent with physical law, is a great problem for the next century.

Many common problems in daily life such as the famous traveling-salesman problem and other resource optimization problems can be reduced to an NP-complete problem. This class of problems possesses the tantalizing property that if an efficient solution to one example can be found, then all such problems possess an efficient solution. Quantum search methodology provides a new perspective with some potential for this notoriously hard but important class of computational problems.

The immense significance of Shor's quantum algorithm for factoring has led to extensive work toward the development of new quantum algorithms for other problems. More generally, characterizing the limitations of efficient problems for quantum computers is just as important as finding new quantum algorithms, because it would allow us to implement classical cryptosystems today that would be impervious to quantum cryptoanalysis should quantum computers ever become practical.

USING A QUANTUM PROCESSOR TO PREDICT THE BEHAVIOR OF COMPLEX QUANTUM SYSTEMS

Richard Feynman pointed out that since everything we experience in our physical world is made from the microscopic building blocks of electrons, nuclei, photons, and so forth, a quantum mechanical description will be essential to understand all of its details. But, he contended, our computers calculate in the classical world and will never be up to the task of such complex quantum calculations. Predicting the behavior of a single qubit can be done with reasonable accuracy by using a classical computer to simulate its dynamics according to the rules of quantum mechanics. However, when many qubits interact, the number of possible states or configurations of the system increases exponentially with the number of qubits, and prediction becomes an impossible task. For example, 30 qubits can be in any of 2^{30} (about a billion) states. Two decades ago a typical desktop computer could solve for the energies and configurations of 10 interacting electrons. Today's conventional computers are more than 100 times as powerful, yet this only allows us to solve a system with two more electrons.

Feynman recognized in 1982 that one quantum system is like any other when measured by its capacity to store and process information, just as information processing with the classical bit is independent of the form of the bit. This means that an atom with two energy levels can be made to act like a photon with two polarization directions, which can be made to act like an electron with two spin orientations, and so on. These quite different physical systems can all be represented as qubits. A quantum simulation is a computation using many qubits of one system type that can be initialized and controlled in the laboratory to simulate an equal number of qubits of another type that cannot be easily controlled. Quantum simulations are special cases of quantum computations and hence may be much easier to implement. For instance, a model-sized device of 30 qubits would be able to perform calculations on quantum systems that would require an array of 1 billion billion values to be represented in a conventional classical computer, far beyond the memory capacity of any of today's computers.

It is increasingly being realized that such quantum simulations can have practical utility. For example, a long-term problem in statistical physics and engineering is finding solutions to lattice spin models. Spin lattices are regular arrays of quantum-mechanical spins whose interactions are limited to short distances. An example is the common permanent magnet (ferromagnet). More complex systems can display "frustration" (see Figure 7-16). Because spin lattice models capture the essential

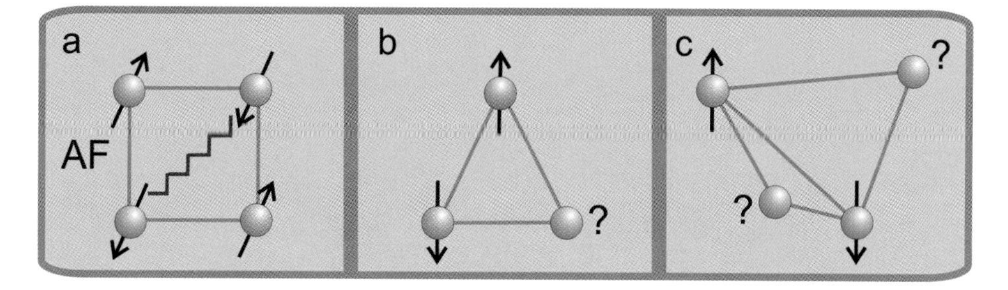

FIGURE 7-16 Frustration in magnetic spin systems. (a) In an antiferromagnetic material, magnetic interactions between atoms cause adjacent spins to have alternate orientations. (b) In the simplest example of spin frustration, given two antialigned spins in an antiferromagnetic system, the state of the third spin is confused. (c) With four spins, there is even more ambiguity. Calculating the ground states of frustrated spin systems and quantum phase transitions in the spin orientations is very difficult using conventional computers. SOURCE: P. Lemmens, Institute for Condensed Matter Physics, Technische Universtät Carolo-Wilhelmina zu Braunschweig, Germany.

physics of more complicated systems, their relevance ranges from commercial materials fabrication to basic problems in condensed matter physics. Another exciting prospect for quantum simulations is to gain insight into the mechanism of high-temperature superconductivity, for which the correct microscopic physics is still unknown. Various models have been proposed, but even simplified versions of these involve solving a highly correlated many-body system. Classical algorithms are not adequate to predict the correct macroscopic behavior of these models. Quantum simulations using trapped ultracold fermionic gases for which a high-temperature superfluid phase has recently been discovered (Chapter 3) may provide critical insight into the mechanism for this high-temperature superconductivity.

There may be a less practical but equally exciting use for quantum simulators—namely, to strengthen our understanding of nature at the subatomic level. For instance, a variety of spin lattice models have been proposed that create low-energy states with a special kind of global property termed topological order. Such theories predict that some physical entities assumed to be fundamental in the Standard Model of particle physics, such as electrons and light, are actually low-energy excitations of a network of spins. If this were actually the way the universe worked then the underlying spin network would be operating at the incredibly small Planck scale, far beyond our level of observation. But the models themselves can be simulated and their predictions tested in a tabletop experiment. Quantum simulators open the door to testing this and other modern theories of quantum cosmology.

LOOKING FORWARD

One of the most tantalizing aspects of quantum information science is that a quantum computer may eventually arise from technology that is completely unknown at present. Whatever the evolution of such devices, experiments that control individual atoms and photons will continue to lead exploration of the bizarre features of quantum mechanical foundations and their application to information processing. After all, the systems currently under study are precisely the "thought experiments" envisioned by Einstein, Bohr, Heisenberg, and the other founding figures of quantum theory over 80 years ago. With the new language of quantum information, we hope to gain more insight into the underlying quantum physical principles, just as the classical theory of information ushered in the advances in physics responsible for our current digital age.

The interdisciplinary nature of this field and the important role played by theory are characteristic trademarks of quantum information. Theory continues to play a key role, even as experimental progress is made toward the quantum processing of information for communication and computational tasks. This exciting and rapidly moving area promises to be a major driver of AMO physics in the next few decades, demanding advances in control of our quantum world that will continue to surprise and amaze.

8

Realizing the Future

The intellectual merit and tremendous discovery potential of the AMO science described in the previous chapters of this report are unmistakable, and AMO technologies are of high value to U.S. needs, including those of commerce, defense, and medicine. The future for this field is enormously promising, both as an engine of fundamental discovery and as a generator of essential technologies that foster progress in other disciplines and in industry. Indeed, in the 12 years that have elapsed since the last decadal survey of AMO science,[1] 15 Nobel prizes have been awarded to researchers in this field.[2] AMO science and technology is in a period of remarkable productivity and importance. Thus, strength in this discipline is an important part of maintaining U.S. leadership in science and technology generally and in the physical sciences in particular.

The purpose of this last chapter is to put these developments in AMO science in the present-day context of science in the United States and the world. Here the committee discusses the budgetary and political outlook for the near and mid-term future, as well as U.S. needs for technically trained people in this very important area. Based on an analysis of future scientific opportunities and of data gathered

[1]National Research Council, *Atomic, Molecular, and Optical Science: An Investment in the Future*, Washington D.C., National Academy Press (1994) (known as the FAMOS report).

[2]Physics Nobel prizes: (1997) S. Chu, C. Cohen-Tannoudji, and W.D. Phillips; (2001) E.A. Cornell, W. Ketterle, and C.E. Wieman; and (2005) R.J. Glauber, J.L. Hall, and T.W. Hänsch. Chemistry Nobel prizes: (1996) R.F. Curl, H. Kroto, and R.E. Smalley; (1998) W. Kohn and J. Pople; and (1999) A. Zewail.

from the federal funding agencies,[3] the committee came to a set of conclusions and recommendations that can guide U.S. policy makers and funding agencies as they distribute precious resources in difficult times. The conclusions are in the final section of this chapter.

THE CURRENT STATUS OF AMO PHYSICS PROGRAM SUPPORT

The key U.S. agencies supporting the field (DOE, NIST, and NSF) have recognized the recent extraordinary achievements in AMO physics and have responded well to the new opportunities that have emerged, with budget increases in real terms (that is, in constant FY2005 dollars) of 26 percent over the last decade (Table 8-1). These increases, which required difficult choices by program staff, are all the more notable given the inflation-adjusted flat budgets for physical science over the same period (as measured by the Consumer Price Index (CPI)).[4] The result of these investments has been an outpouring of excellent science. Indeed, the AMO example is an interesting case study for the benefits of federal investments in science generally: Good funding of excellent science leads to an excellent return on investment, which in turn leads to new opportunities.

In contrast, there has been a worrisome trend in the Department of Defense (DOD) science agencies to cut back on funds for research generally and for basic research in AMO physics in particular. This policy, which reflects heightened priorities for national security and homeland defense, cannot be justified as a wise or effective long-term strategy if the goal is maximizing the overall strength of the nation. This judgment on the importance of S&T to our overall national security (broadly defined) has been asserted many times over the years in a number of major reports,[5]

---

[3]The committee prepared a questionnaire for the federal agencies that support AMO research in order to collect information on the current trends in AMO science, personnel, training, and funding. The agencies involved are the Air Force Office of Scientific Research (AFOSR), the Army Research Office (ARO), the Defense Advanced Research Projects Agency (DARPA), the Department of Energy (DOE), the National Aeronautics and Space Administration (NASA), the National Science Foundation (NSF), the National Institute of Standards and Technology (NIST), and the Office of Naval Research (ONR). Appendix A contains the questionnaire and Appendixes B-F contain the responses.

[4]For a guide to R&D funding data by science and engineering discipline, see the Web site of the American Association for the Advancement of Science at <http://www.aaas.org/spp/rd/guidisc.htm>.

[5]See the following reports: House Committee on Science, *Unlocking Our Future: Toward a New National Science Policy* (1998), available at <http://www.house.gov/science/science_policy_report.htm>, accessed June 2006; *Before It's Too Late* (Glenn Commission) (2000), available at <http://www.ed.gov/inits/Math/glenn/report.pdf>, accessed June 2006; The United States Commission on National Security/21st Century, *Road Map for National Security: Imperative for Change* (2001), also known as the Hart-Rudman Report, available at <http://www.fas.org/man/docs/nwc/phaseiii.pdf>, accessed June 2006; National Science Foundation, *The Science and Engineering Workforce: Realizing America's Potential* (2003), available at <http://www.nber.org/~sewp/>, accessed June 2006; U.S. Domestic

TABLE 8-1 Funding Histories in AMO Science (1996-2005) (millions of dollars)

	DOE		NIST	NSF		DOE/NIST/NSF[a]			
Year	Centers	Total	Total	Centers	Total	As-Spent Total	Deflator	FY2005 $	DOD/NASA[b]
1996	5.81	9.93	44.90	2.4	17.45	72.3	0.800	90.3	See notes
1997	5.70	9.93	46.80	2.4	17.54	74.3	0.820	90.6	below and
1998	6.01	9.93	48.10	2.4	19.59	77.6	0.843	92.1	Appendix B
1999	6.01	11.02	50.30	2.4	20.79	82.1	0.861	95.4	
2000	5.94	10.77	51.40	3.75	21.74	83.9	0.890	94.3	
2001	6.34	11.43	55.50	7.25	27.10	94.0	0.916	102.7	
2002	6.95	11.82	61.30	7.22	28.16	101.3	0.930	108.9	
2003	7.30	13.38	65.00	7.33	30.95	109.3	0.951	115.0	
2004	7.54	13.88	66.90	7.45	26.62	107.4	0.977	109.9	
2005	7.68	16.63	70.00	7.56	27.24	113.9	1.000	113.9	

[a]Notes on DOE, NIST, and NSF

1. Table includes only information from NSF/PHY, DOE/BES/AMOS, and NIST (see Appendixes B, D, and E).
2. Indicates that there has been ~26 percent real growth in AMO funding over the decade from these sources.
3. In constant FY2005 dollars, DOE has grown by ~34 percent, NIST by ~25 percent, and NSF by ~25 percent.
4. Other parts of DOE and NSF support AMO science also. Sometimes this support is very large (e.g., construction of LCLS by DOE/BES). These funds are not included here.
5. Amounts to theory: DOE, ~25 percent; NIST, ~7 percent; and NSF, ~17 percent.
6. AMO funds to universities: DOE, ~60 percent; NIST, ~10 percent; NSF, 100 percent. Remainders stay in-house.
7. Many investigators in AMO science have multiple means of support.

[b]Notes on DOD agencies (AFOSR, ARO, DARPA, ONR) and NASA

1. The data supplied to the committee by the DOD agencies and by NASA were mostly anecdotal.
2. The DOD agencies have a long history of substantial funding to AMO science, though this declined over the last decade or so. In FY2005, support levels were approximately as follows: AFOSR, $5.5 million; ARO, $5 million; DARPA, >$30 million; and ONR, $5 million. Thus these agencies added more than $45 million to the above total (see Appendix A). Recent funding also been at about this level, though a detailed accounting is difficult because of program variability and high interdisciplinarity. Yearly funding data were not supplied to the committee.
3. Amounts to universities are AFOSR, ~75 percent; ARO, 100 percent; and ONR, ~67 percent. DARPA's funds also go predominantly to universities.
4. Amounts to theory are AFOSR, ~10 percent; ARO, ~20 percent; DARPA, ~0 percent; and ONR, ~33 percent.

including, recently, *Rising Above the Gathering Storm*:[6]

> The United States takes deserved pride in the vitality of its economy, which forms the foundation of our high quality of life, our national security, and our hope that our children and grandchildren will inherit ever-greater opportunities. That vitality is derived in large part from the productivity of well-trained people and the steady stream of scientific and technical innovations they produce. Without high-quality,

Policy Council, *American Competitiveness Initiative* (2006), available at <http://www.whitehouse.gov/stateoftheunion/2006/aci/aci06-booklet.pdf>, accessed June 2006.

[6]NAS/NAE/IOM, *Rising Above the Gathering Storm: Energizing and Employing America for a Brighter Economic Future*, Washington, D.C.: The National Academies Press (2007), p. 1.

knowledge-intensive jobs and the innovative enterprises that lead to discovery and new technology, our economy will suffer and our people will face a lower standard of living. Economic studies conducted before the information-technology revolution have shown that even then as much as 85 percent of measured growth in U.S. income per capita is due to technological change.

Today, Americans are feeling the gradual and subtle effects of globalization that challenge the economic and strategic leadership that the United States has enjoyed since World War II. A substantial portion of our workforce finds itself in direct competition for jobs with lower-wage workers around the globe, and leading-edge scientific and engineering work is being accomplished in many parts of the world. Thanks to globalization, driven by modern communications and other advances, workers in virtually every sector must now face competitors who live just a mouse-click away in Ireland, Finland, China, India, or dozens of other nations whose economies are growing.

In view of these concerns,

the National Academies was asked by Senator Lamar Alexander and Senator Jeff Bingaman of the Committee on Energy and Natural Resources, with endorsement by Representatives Sherwood Boehlert and Bart Gordon of the House Committee on Science, to respond to the following questions:

> What are the top 10 actions, in priority order, that federal policy-makers could take to enhance the science and technology enterprise so that the United States can successfully compete, prosper, and be secure in the global community of the 21st Century?
>
> What strategy, with several concrete steps, could be used to implement each of those actions?

The *Gathering Storm* report responds with four major recommendations (and a number of secondary ones):

Recommendation A: Increase America's talent pool by vastly improving K-12 science and mathematics education.

Recommendation B: Sustain and strengthen the nation's traditional commitment to long-term basic research that has the potential to be transformational to maintain the flow of new ideas that fuel the economy, provide security, and enhance the quality of life.

Recommendation C: Make the United States the most attractive setting in which to study and perform research so that we can develop, recruit, and retain the best and brightest students, scientists, and engineers from within the United States and throughout the world.

Recommendation D: Ensure that the United States is the premier place in the world to innovate; invest in downstream activities such as manufacturing and marketing; and create high-paying jobs that are based on innovation by modernizing the patent system, realigning tax policies to encourage innovation, and ensuring affordable broadband access.

Perhaps the most prominent of the secondary recommendations in the *Gathering Storm* is the call to establish a special new focus on research into alternative sources of energy (via the creation of a new agency, DARPA-E)—to relieve our "addiction to oil," as President Bush framed the issue in his 2006 State of the Union address. In that speech, President Bush commented on many of the themes that appear in the *Gathering Storm* and announced the American Competitiveness Initiative to start in FY2007, which includes an advanced energy initiative comprising a number of proposed activities across a broad front of energy research. The committee believes that strong federal stewardship of the opportunities and challenges in the physical sciences generally and in AMO science specifically will be an important element of these initiatives. AMO science addresses directly the report's major recommendations (C and D, above). AMO science also provides critical research paths that will be necessary to meet the challenges in attaining future energy security.

These recommendations assume even greater urgency because over the last decade, U.S. funding for the physical sciences overall has just kept pace with inflation.[7] However, many observers believe that the cost of doing research rises significantly faster than the CPI would indicate—in particular due to the dramatic rise in the capability of new instrumentation. Increasing computational power and new capabilities in electronics allow building instruments that were simply unimaginable only 5 or 10 years ago. While this progress opens rich and productive new avenues for research, following them is very expensive. Agency budgets are simply not rising fast enough to keep up.

While federal spending for AMO science has seen some real growth (see Table 8-1 and Appendix B) over the last decade, investments in the life sciences over the same period—especially in medicine—have shown dramatic growth (see Figure 8-1). There have been many calls from Congress (by individual members, by congressional committees, in congressionally sponsored studies), from the business community, and from leaders in the life sciences community—including the director of the National Institutes of Health (NIH)—for the same kind of increases for the physical sciences as have been allocated for the biomedical sciences. But despite broad bipartisan congressional support for doubling the NSF budget and an authorization act that allows doing so, this doubling has not occurred. At the DOE Office of Science, budgets have declined in constant FY2004 dollars over the last 12 years, though there has been growth in the Office of Basic Energy Sciences.

Unfortunately, because of very significant pressures to constrain federal spending, federal allocations for physical science have remained static at best and at

[7]For a guide to R&D funding data by science and engineering discipline, see the Web site of the American Association for the Advancement of Science at <http://www.aaas.org/spp/rd/guidisc.htm>.

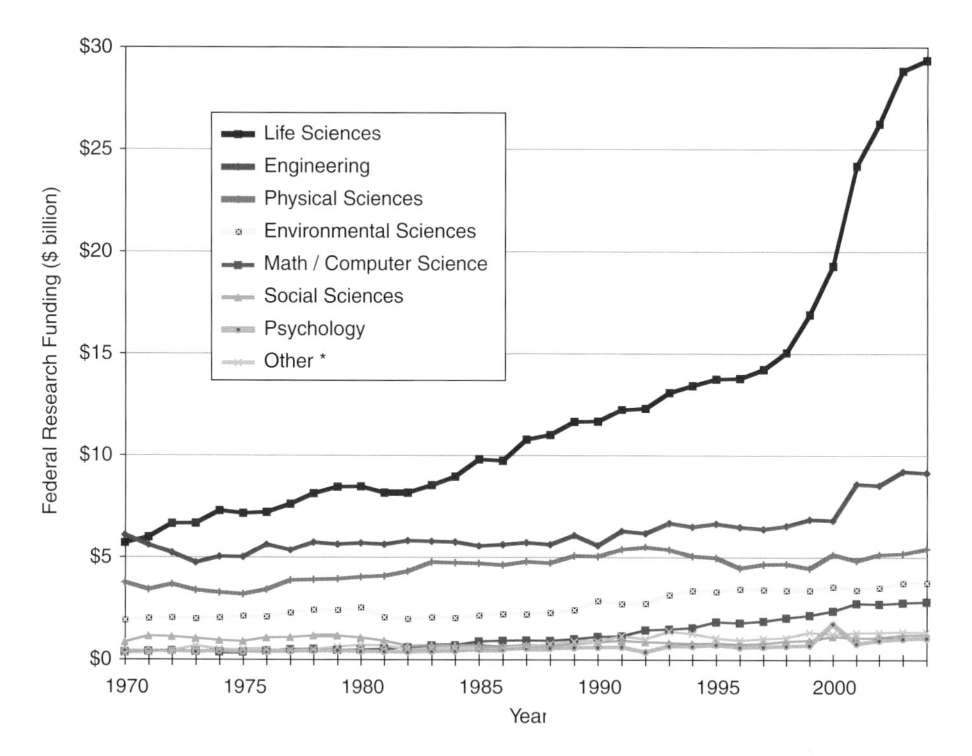

FIGURE 8-1 Trends in federal research by discipline, FY1970-2004, obligations in billions of constant FY2004 dollars. Data are based on National Science Foundation, Federal Funds for Research and Development FY2002, 2003, and 2004. FY2003 and 2004 data are preliminary. SOURCE: American Association for the Advancement of Science. NOTE: "Other" includes research not classified (basic research and applied research; excludes development and R&D facilities).

times decreased over the medium term. The committee believes that continuing this policy would lead to a dangerous outcome for the United States. However, the increases sought in the President's 2007 budget request are reassuring. There are very few dollars that the government spends that bring more added value for our country than the funds spent on scientific research and on the education of our future scientific workforce.

MAINTAINING U.S. LEADERSHIP IN A CRITICAL AREA OF SCIENCE AND TECHNOLOGY

Countries around the world recognize the enormous value of AMO science and technology. The international AMO scene is highly competitive, and nations in Europe and the Far East are making substantial investments in this area (see Appendix C). Of the 15 Nobel laureates mentioned in footnote 2, 5 either did their ground-breaking research in Europe or the Middle East or were trained there. More generally, foreign investment in science, as a percentage of GDP, in many cases surpasses the U.S. investment. Among the countries of the Organisation for Economic Co-operation and Development (OECD)[8] countries, the United States ranked fifth in the ratio of total R&D to GDP and sixth with the inclusion of Israel.[9]

As the *Gathering Storm* report stresses, in recent years it has become very clear that maintaining U.S. leadership in physical science depends on more than simply money. Since at least the 1980s, the United States has benefited from a large influx of foreign nationals to fill the ranks of the scientific and engineering workforce. Indeed, according to a recent survey by the Council of Graduate Schools,[10] engineering and physical science have been the leading attractors among all fields of graduate study for foreign-born students. Figure 8-2 shows that in recent years only about 52 percent of the Ph.D.s granted in the United States have been to U.S.-born students. AMO science has benefited greatly from the influx of talented foreign-born students.

Now that there are excellent opportunities for these talented people in their home countries, they are either staying there to start with or training in the United States and then returning home. This has left the United States in a potentially vulnerable position, since recent history shows that American-born students are generally making career choices outside science. Without remediation, the United States could fall into a capability gap. The National Science Board recently commented on this situation:[11]

> Every two years the National Science Board supervises the collection of a very broad set of data trends in science and technology in the United States, which it publishes as *Science and Engineering Indicators* (*Indicators*). In preparing *Indicators 2004*, we have observed a troubling decline in the number of U.S. citizens who are training to become scientists and engineers, whereas the number of jobs requiring science and engineering (S&E) training continues to grow. Our recently published report entitled *The Science and Engineering Workforce/Realizing America's Potential* (NSB 03-69, 2003) comes to a similar conclusion. These trends threaten the economic welfare and security of our country.

[8]Current members of the OECD are Australia, Austria, Belgium, Canada, Czech Republic, Denmark, Finland, France, Germany, Greece, Hungary, Iceland, Ireland, Italy, Japan, Korea, Luxembourg, Mexico, Netherlands, New Zealand, Norway, Poland, Portugal, Slovak Republic, Spain, Sweden, Switzerland, Turkey, United Kingdom, and United States.

[9]See <http://www.nsf.gov/statistics/seind04/>.

[10]See <http://www.cgsnet.org/>.

[11]See *A Companion to Science and Engineering Indicators 2004*, available at <http://www.nsf.gov/statistics/nsb0407/>.

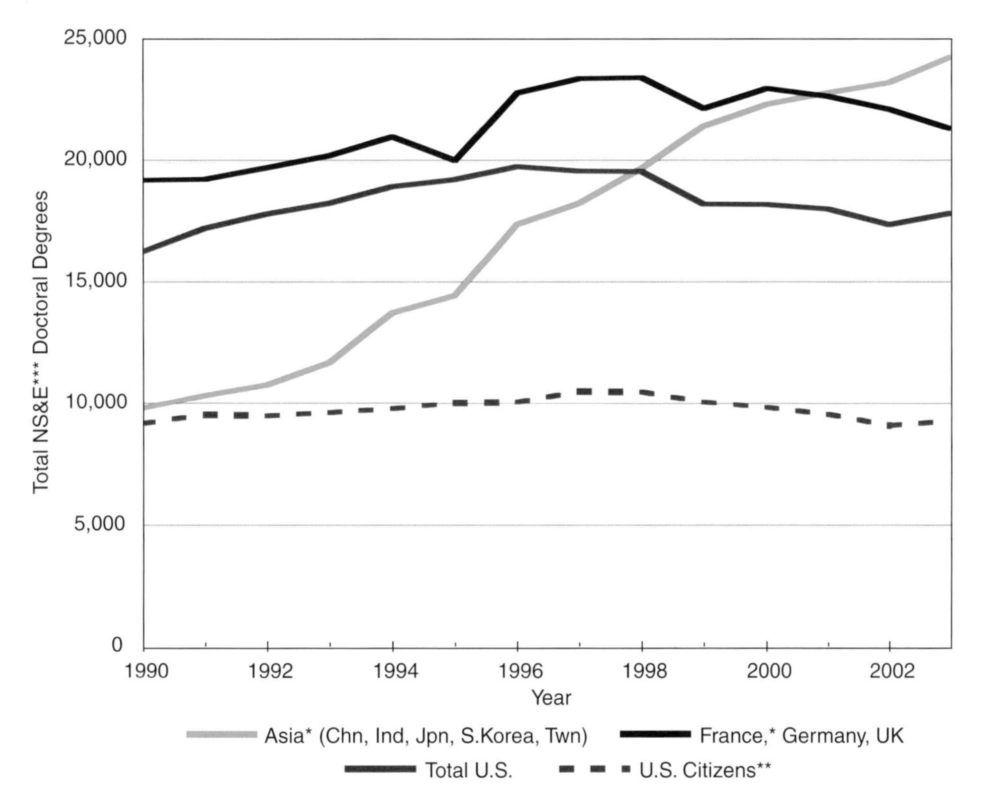

FIGURE 8-2 Natural science and engineering Ph.D.s granted in the United States compared to the major European nations and to Asia. About 52 percent of U.S.-granted Ph.D.s are awarded to American-born students. Data drawn from 2006 NSF Science and Engineering Indicators. SOURCE: American Physical Society, Washington office. NOTES: *Some data were not available, so as a conservative estimate, prior-year data were used. The 2001 China data were used for the 2002 and 2003 values; 2001 Taiwan value for 2002 value; 2000 and 2002 South Korea data for 2001 and 2003, respectively; 1999 France value for 2000-2003. **U.S. institutions only. ***NS&E degrees include natural (physical, biological, Earth, atmospheric, and ocean sciences), agricultural, and computer sciences, mathematics, and engineering.

If the trends identified in *Indicators 2004* continue undeterred, three things will happen. The number of jobs in the U.S. economy that require science and engineering training will grow; the number of U.S. citizens prepared for those jobs will, at best, be level; and the availability of people from other countries who have science and engineering training will decline, either because of limits to entry imposed by U.S. national security restrictions or because of intense global competition for people with these skills. The United States has always depended on the inventiveness of its people in order to compete in the world marketplace. Now, preparation of the S&E workforce is a vital arena for national competitiveness.

> Even if action is taken today to change these trends, the reversal is 10 to 20 years away. The students entering the science and engineering workforce in 2004 with advanced degrees decided to take the necessary math courses to enable this career path when they were in middle school, up to 14 years ago. The students making that same decision in middle school today won't complete advanced training for science and engineering occupations until 2018 or 2020. If action is not taken now to change these trends, we could reach 2020 and find that the ability of U.S. research and education institutions to regenerate has been damaged and that their preeminence has been lost to other areas of the world.

How will the United States respond to this overall situation? How will our country maintain its scientific leadership in AMO science and technologies and in other S&T areas, too, in the face of these budgetary and personnel pressures? Clearly it is a matter of strategic national importance. Because these are very serious issues, it is not surprising to find that they were addressed not only in the January 2006 State of the Union address, but also in the White House budget priorities and guidelines for 2007, in statements from many members of Congress, and in the plans of the funding agencies. The White House guidance on budget priorities for FY2007 contained the following language:[12]

> Investments in the physical sciences likely to lead to or enable new discoveries about nature or strengthen national economic competitiveness continue to be important. Priority will be given to research that aims to close significant gaps in the fundamental physical understanding of phenomena that promise significant new technologies with broad societal impact. High-temperature and organic superconductors, molecular electronics, wide band-gap and photonic materials, thin magnetic films, and quantum condensates are examples of novel atomic and molecular-level systems with such gaps where coherent control holds great potential.

The committee welcomes the attention that these issues are receiving in the government. The President's American Competitiveness Initiative and other aspects of the FY2007 budget request are dramatic further steps. This initiative has a number of important goals, including increases in federal investments in critical research areas in order to ensure that the United States continues to lead the world in opportunity and innovation and to provide American children with a strong foundation in math and science. In his remarks, President Bush asked Congress to double the federal commitment to the most critical basic research programs in the physical sciences over 10 years, including, specifically, a doubling of the sum of funding available to NIST, the NSF, and DOE's Office of Science.

[12]Available at <http://www.whitehouse.gov/OMB/memoranda/fy2005/m05-18.pdf>.

PLANNING FOR FUTURE U.S. LEADERSHIP IN AMO SCIENCE

What must be done to realize the future promise of AMO science and technology and to capitalize on AMO's many potential contributions to overall U.S. interests and national needs? The answers revolve around four major goals for the nation:

- We must train the next generation of AMO scientists.
- We must fund the field adequately in order to realize its discovery potential.
- We must be alert to new and more efficient ways of doing these things: new training modalities and new operational structures (centers, multi-investigator grants, formation of international teams, stronger cooperation between government and the private sector, and so on).
- We must deal effectively with known or potential problems—for example, stimulating the growth of our own national workforce while continuing to attract foreign talent; continuing a steady flow of funding resources, with increases where possible; compensating for lost or shifting funding at some of the funding agencies; building links to the private sector where possible; achieving a good balance between mission-oriented and so-called "blue-sky" research; and responding to international competition.

The committee prepared a questionnaire for the federal agencies that support AMO research in order to collect information on the current trends in AMO science, personnel, training, and funding.[13] In addition to those agencies the committee polled, there are other federal agencies that support research in AMO science. For example, at NIH AMO science plays a substantial interdisciplinary role in programs within the National Institute for General Medical Sciences and the National Institute for Biomedical Imaging and Bioengineering, but there is no grants program specific to AMO efforts. The National Security Agency (NSA) supports work in quantum information technology and possibly other parts of AMO science, but the committee did not attempt to gather information about it. Nor was information gathered from the Disruptive Technologies Office (DTO) or the National Nuclear Security Administration.

The responses received fall into four broad categories:

- Information about the intellectual portfolio of the research currently supported by the federal government,
- Information about its funding,

[13]See footnote 3.

- Information about the workforce and the training of the next generation of scientists, and
- Information about new operational modalities for carrying out the research.

Each of these topics is discussed below. In view of the great diversity in these programs, the discussion is presented in some depth. The discussion also includes other issues that are important in shaping an effective national strategy, including the increasing importance of international competition in AMO science and important logistical issues emerging here at home.

Intellectual Outlines of Research Currently Supported

Appendix D contains the information the committee received about the current research profiles of the funding agencies. With the exception of the NSF, all of the agencies supporting AMO science are mission-oriented. Because proposals to NSF can be unconstrained in scientific content, the community sends in what it believes to be the most promising science. Applicants to the other AMO programs propose research in support of the agency mission and program goals. NSF judges the proposals it receives entirely on the basis of peer review, and DOE and NASA also use peer review to a great extent. Other agencies allocate funds based largely or entirely on the decisions of internal program staff.

The committee concludes that all the agencies, in spite of a significant mission orientation, strive to create dynamic, high-quality portfolios that advance scientific and technical knowledge as they meet those mission requirements. There is, however, considerable concern that the DOD agencies have decreased support for research generally and for basic research in particular (see below).

The AMO research profiles of federal agencies have shifted to newer areas such as intense field and ultrafast x-ray science, cooperative phenomena, ultracold atoms and molecules, precision measurements, nanoscale science, and quantum optics. The highly interdisciplinary, central role of AMO science is evident—the work clearly is of great interest to physics as a whole.

There has been a substantial shift of support away from the more traditional core areas of AMO physics, such as atomic and molecular collision physics and spectroscopy. However, a large segment of this work is critical to many applications, such as plasma processing, lighting, gaseous electronics, and sensor development. This work is also critical to forefront astronomical research, especially as newer ground-based and space-based instruments are being developed with greater photon sensitivities and higher spectroscopic resolution (see Appendix D and Box D-1).

One of the great strengths of U.S. research is its diversity. Nowhere is this clearer than in AMO science, with its very wide intellectual horizon, its variety of funding sources, and its several modalities available for accomplishing the work.

This discussion of the research profile of AMO science is not complete without a comment about the essential role played by theory. In any discipline where the frontiers are traversed so rapidly, theoretical research is essential—not only to understand what has been observed but also to predict what might be seen next and thereby guide future work. This is especially true in AMO science, where researchers are developing and pursuing dramatic new areas such as ultra-high-field physics, the physics of "sparse" condensed matter systems, the study of the fundamentals of quantum mechanics, and the frontiers of quantum information.

The committee therefore notes with concern that support for AMO theory at NSF has not progressed over the last decade, even while support for experiment has been relatively strong. The size of a single-investigator grant in theory at NSF is far below that at the other agencies (see the following section, Table 8-2, and

TABLE 8-2 Demographics, Success Rates, Turnover, Average Grant Sizes, FY2005

	AFOSR	ARO	ONR	DOE	NASA[a]	NIST[c]	NSF
Awards[b]	~24	~20	~30	~55	~45	~5	~135
Senior[d]	~30	~40	~30	~83	~45	~160	~145
Postdoc	~30	~30	NA	~50	NA	~20	~50
Ph.D.	~6	~10	NA	~30	NA	~5	~40
Graduate students	~30	~50	NA	~100	NA	~30	~200
Undergraduate students	NA	NA	NA	NA	NA	~20	~115
Women (%)[e]	~10	~10	NA	NA	NA	~16	~10
Minorities (%)[e]	~7	<5	NA	NA	NA	<5	~2
Average grant size (thousand $)							
Experimental	~125	~140	~150	~137	~150	~100[f]	~135
Theory	~80	~80	~125	~104	NA	~100[f]	~60
Success rate (%)[g]	~30	~40	~40	~34	~35	N/A	~43
Turnover (%)[h]	~5	~15	~15	~10	NA	N/A	~7

NOTE: Personnel numbers include both experimental and theory. Owing to the high variability and interdisciplinarity of its programs from year to year, DARPA was not included in this table.

[a]NASA data include the Laboratory Astrophysics and Planetary Sciences programs.

[b]Awards from NIST are solicited by the agency.

[c]Number of grants in place.

[d]Number of senior investigators supported.

[e]Percentages of the supported investigator pool represented by women or minorities.

[f]Funding per scientist at NIST labs not including overhead, depreciation, and operating expenses.

[g]Percentage of proposals funded out of those submitted; many ideas for proposals are "declined" before a submission.

[h]The rate at which new people enter the program.

Appendix B). This policy of low average grant size seems to be based not on any reasonable assessment of the needs of a single investigator conducting research in a university environment, but rather on the need to maintain a reasonable number of active theory investigators within an essentially fixed program budget. NSF also supports critically important work at the Institute for Theoretical Atomic, Molecular and Optical Physics (ITAMP), although this funding level has also not increased over the last decade. A limited amount of work in AMO theory is also carried out at the Kavli Institute for Theoretical Physics in Santa Barbara (which is partially supported by NSF) as a part of its workshop series. The committee believes that this pattern of funding erosion in theoretical AMO physics at NSF is not an effective way to realize the full contribution that theory could make to the compelling science opportunities detailed in this report. Theoretical work is also carried out within the DOE and NIST programs, and to a lesser extent in the DOD science agencies. In all these cases the support per investigator is much more robust.

Finally, the committee notes that there are many separate programs throughout the government for funding quantum computing research. Perhaps only 25-35 percent of that effort is in the AMO community, but given the size of the quantum computing funding, even that fraction is a significant amount. The key sponsors are ARO, AFOSR, DARPA, DTO, NSF, and NSA.

Information About Funding

Appendix B and Tables 8-1 and 8-2 contain data the committee received on agency budgets.[14] As noted above, in contrast to the support for physical science in general, there was a significant increase in AMO support (about 26 percent in FY2005 dollars) over the last decade at DOE, NIST, and NSF, with much of the increases at these agencies going to the creation or expansion of activities at centers.

In contrast, a recent NRC report[15] found that "in real terms the resources provided for Department of Defense basic research have declined substantially over the past decade." This is particularly true for ONR, which had a strong tradition, until recently, of supporting some innovative and long-term AMO research. These so-called 6.1 funds in the DOD budget appear to require continuous justification to be included as a part of DOD agency missions, which have become increasingly

[14]It was not possible to compare the data this committee received with the data collected for the 1994 FAMOS report, since the research boundaries of the field were defined differently in the two reports. The FAMOS committee appears to have interpreted the field more broadly and, in addition, received more information from the agencies.

[15]NRC, *Assessment of Department of Defense Basic Research*, Washington D.C., The National Academies Press (2005), Finding 15.

oriented to short- and medium-term goals. The resultant reductions in 6.1-supported research have had a substantial impact in the university community, which tends to focus on longer-term basic research.

It is contended by some that this problem arises not only from a direct reduction of available 6.1 funds, but also from a relabeling of previous 6.2 and 6.3 work into the 6.1 category. However, NRC's FAMOS report found little evidence for this contention. On the other hand, Finding 9 of the report goes on to assert that "generated by important near-term Department of Defense needs and by limitations in available resources, there is significant pressure to focus DOD basic research more narrowly in support of more specific needs." And this is taking place at a time when, according to Finding 14, "the breadth and depth of the sciences and technologies essential to the Department of Defense mission have greatly expanded over the past decade." And finally, from Finding 8: "A recent trend in basic research emphasis within the Department of Defense has led to a reduced effort in unfettered exploration, which historically has been a critical enabler of the most important breakthroughs in military capabilities." A compelling example of such a breakthrough in research on atomic clocks that was supported by ONR is described in Boxes 2-3 and 7-3 of this report. This situation has led to much increased proposal pressure at agencies that do support fundamental work. For example, AFOSR is now seeing more proposals from people who used to be funded by ONR and NASA.

With the possible exception of AFOSR, the DOD agency programs seem to vary considerably from year to year in terms of available funds. Owing also to the high degree of multidisciplinary work supported, it is difficult to estimate accurately the funding available each year to AMO science from the DOD agencies, especially at DARPA. From the data gathered for this study, an estimate is $45 million per year—but with a significant uncertainty on this number. About $30 million of this is from DARPA alone. However, the data supplied to us by the DOD agencies and by NASA were mostly anecdotal. Year-by-year funding trends were not made available to us.

It is a testament to the vitality of the field that all of the federal agencies report that the number of very high quality proposals is much larger than the available funds can support. And based on funding requests by grantees currently under review, most of the agencies report that grant sizes in experimental AMO physics could usefully be about 30-50 percent higher—though it is certainly true that grant sizes for the same level of effort will vary depending on the institution, its location, salaries, indirect cost rates, and so on. But in cases where budgets have remained flat, there have been serious losses in the purchasing power of AMO scientists due to regular CPI inflation and the instrumentation capability inflation mentioned above. Maintaining flat levels of effort means that young people are not coming into the field at a higher rate. More than that, the incremental investment that these additional funds represent would ensure that the government's start-up investment yields as high a scientific return as possible.

AMO scientists often receive support from multiple sources simultaneously, including support from their home institution. The committee did not attempt to quantify the degree to which this occurs, or what it may mean in terms of total support, on average, for an AMO scientist. Nor were data gathered on possible funding from the private sector. However, it is widely recognized that funding for basic research at most of the country's leading private industrial research laboratories has all but vanished. Today there is significant interest in the industrial sector in laser physics across an enormous range of activity, but this interest is focused almost entirely on commercial R&D development and hardly at all on fundamental physics. A possible exception is in quantum information science and quantum computing, where companies like IBM and Microsoft have very active research programs in laser physics, condensed matter physics, and other areas germane to those topics.

Information About People

Appendix E contains data the committee received on the makeup of the AMO community of scientists and students. Generally speaking, the number of people supported in the field has not increased dramatically over the last decade, although requests for support have increased. As a measure of this, the committee notes that the membership of the Division of Atomic, Molecular and Optical Physics (DAMOP) of the American Physical Society (APS) has remained relatively constant for more than a decade at ~2,550, not changing from year to year by more than about 5 percent. AMO physicists account for about 6 percent of the total APS membership, and DAMOP is its third largest division. The APS also has a Division of Laser Science, which comprises about 1,350 members, though its membership overlaps very significantly with DAMOP's, at least in optical science. Another measure of the strength of AMO science and technology is the membership of the Optical Society of America, which today is about 14,000, the large majority of whom represent the engineering community and find their employment in government or private industry.

The excitement of the field and the superb training that it affords have attracted excellent people to it. Of the DAMOP membership, about 700 are students. The number of foreign graduate students, postdocs, and senior investigators remains high. Although physics departments are now reporting increased numbers of Americans majoring in physics, the committee could not determine whether or not this trend extends to AMO science.

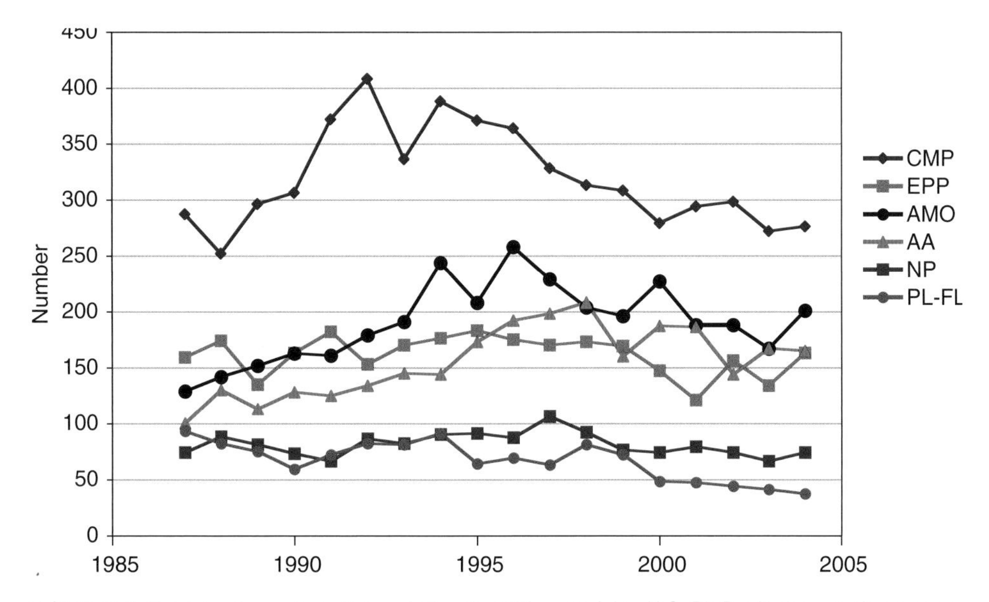

FIGURE 8-3 Total numbers of students of all nationalities receiving U.S. Ph.D.s in the prominent areas of physics, as a function of year. CMP, condensed matter physics; EPP, elementary particle physics; AMO, atomic, molecular, and optical; AA, astronomy and astrophysics; NP, nuclear physics; PL-FL, plasma and fluid dynamics. Total Ph.D.s/year in physics ranged from lows of about 1,240 in 1987 and 2003 to a high of 1,692 in 1994. SOURCE: Data from American Institute of Physics, Statistical Research Center, and the National Science Foundation.

Graduate students and postdocs in AMO science appear to respond very positively to the fact that, for the most part, experiments are still small enough that projects can be conducted by individuals or small groups. Students are therefore able to manage their entire projects nearly on their own, experiencing in full the diversity of the effort—an extremely valuable and highly sought-after training experience.

American Institute of Physics/NSF data (see Figure 8-3) indicate that over the last decade the United States produced about 1,200 Ph.D.s in physics and astronomy each year, of whom about 200 were in AMO science. This number is exceeded only by the condensed matter physics community. Per dollar spent, the AMO community is clearly quite productive in terms of training its next generation.

Information About New Modalities

Appendix F contains information from the agencies about new modalities for doing research. Like other very active areas of science, AMO has evolved substantially over the years in terms of ways the science is carried out. For example, there has been a steady move toward the creation of center activities, supplanting somewhat the traditional emphasis within AMO science on the single-investigator mode of research. The center approach is chosen sometimes so as to assemble in one place a critical mass of research activity capable of generating new ideas in a given area and capitalizing on them more quickly. At times, the nucleation occurs at a new facility such as the Linac Coherent Light Source (LCLS) at Stanford. Finally, the increasing interdisciplinarity of AMO science produces a natural tendency toward collaborative activity at centers where scientists with the requisite expertise can gather and cooperate. This latter point has been explicitly recognized by the DOD agencies that created the Multidisciplinary University Research Initiative program. These awards, which last up to 5 years and give up to $1 million per year, have allowed the agencies to pursue their research objectives very successfully (see Appendix F for more details).

With the creation of its Center for Ultracold Atoms and its Frontiers of Optical Coherent and Ultrafast Science (FOCUS) center, NSF has chosen two very exciting areas of AMO science for robust support. FOCUS builds on the earlier establishment, and evolution, of the Center for Ultrafast Optics, an NSF Science and Technology Center. ITAMP has most of the attributes of a center (visitor's programs, workshops, and a postdoc program) but is not currently funded at the level of the Physics Frontiers Centers program.

Perhaps the most dramatic example of excellent scientific productivity coming out of the "centers" concept is JILA at Boulder, Colorado. Jointly supported by NIST and NSF, JILA has been the undeniable world leader in many areas of quantum optics for over a decade. Its work garnered multiple Nobel prizes in that time. Indeed, all of the AMO work that NIST supports is done within the "centers" concept—at JILA, at the Time and Frequency Laboratory in Boulder, and in Gaithersburg, Maryland.

Almost half (47 percent) of DOE's Basic Energy Sciences (BES) AMO portfolio of research is carried out at centers. Four of the five centers are at national laboratories and one is at Kansas State University. Two centers were new in 2005: the Photon Ultrafast Laser Science and Engineering center is colocated at SLAC with the LCLS and is funded by the Materials Science Program in BES; and the Berkeley Ultrafast Center is located at Lawrence Berkeley National Laboratory.

As described in Appendix F, there is a growing number of large-scale facilities available in this country for research in AMO science (and many smaller ones as well). These include several synchrotron light sources operated by DOE/BES—the Advanced Photon Source at Argonne, the Advanced Light Source at Berkeley, the National Synchrotron Light Source at Brookhaven, and the Stanford Synchrotron Radiation Laboratory—as well as several others at universities. In general, AMO physics accounts for a relatively small fraction of the research portfolios at these facilities. At new facilities, the emphasis is on fourth-generation light sources, by which is meant the development of x-ray free-electron lasers—in particular, the LCLS at SLAC. There are also three very-high-intensity laser systems (OMEGA at Rochester and JANUSP and the National Ignition Facility, both at Livermore). By creating these new research modalities, the federal science agencies have realized a dramatically positive payoff. It is an approach that has also been adopted by the rest of the world.

Foreign Competition

As is already clear from the preceding facilities discussion and as is amplified in Appendix C, the United States is facing intense competition from abroad in AMO science and areas related to it. The same is true for areas relating AMO science and technology to national security. As this worldwide interest attests, there is every indication that AMO science will be an essential frontier of science for years to come. In the words of John Marburger, President Bush's science advisor, "It is not a question of responding to a threat, it is a question of maintaining leadership" in a science of great importance to U.S. scientific, commercial, health, and strategic interests.

The activities going on in AMO science in the rest of the world are as creative, diverse, and robust as the U.S. effort. While it is certainly true that no single country can yet match the overall U.S. program, the sum of what is going on in Europe certainly does. The European Union's support for research on a continental scale, combined with the national efforts of its member states, means that Europe is well-positioned to compete in research efforts of all kinds. Countries like Germany, with a population of 82 million, and Austria, 8 million, have made investments in AMO science that in some cases well surpass U.S. investments even at our most effective and well-funded laboratories (see Box 8-1). The United Kingdom and France, with populations of 60 million each, are not far behind.

One way to measure emerging trends in international competition in research is to look at how the number of scientific publications from other countries has grown over the past two decades. Figure 8-4 shows some publication data from two of the world's premier physics journals, *The Physical Review* and *Physical Review Letters.* The relative position of the United States has steadily slipped: In 1990, U.S. scientists accounted for half the submissions, but by 2004 they accounted for only 25 percent of the total. The same trend is visible in AMO science by itself.

Finally, the committee notes the existence in Europe and Asia of research facilities that either do not exist in the United States (such as heavy-ion storage rings) or that are essentially the equal of U.S. facilities (such as the synchrotron light sources, laser facilities, and free-electron lasers). In terms of strategic resources, Europe's

BOX 8-1
AMO Science in Austria and Germany

In 2003, the Austrian Academy of Sciences founded the Institute for Quantum Optics and Quantum Information. This institute, located in state-of-the art facilities in Innsbruck and Vienna, is composed of four research groups in experimental and theoretical AMO physics. Following the model of the Max-Planck Institutes in Germany, and in particular the Max-Planck Institute for Quantum Optics in Garching (an equally outstanding institute), the institute's goal is to secure a leading role for Austrian science in the fields of quantum optics and quantum information. As a result of this conscious prioritizing and concentrated investment, it has succeeded in doing so.

proposed global positioning satellite system, Galileo, and Russia's existing one, GLONASS, are fully competitive with the U.S. Global Positioning System.

Logistical Issues in the United States

In the aftermath of the 9/11 terrorist attacks, policies across a very broad spectrum were put in place to close existing security holes and to anticipate new ones. Some of these policies, if implemented as originally envisioned, would have had a significant dampening impact on the traditional ways that scientific research has been carried out in the United States and in a larger sense may have worked against U.S. interests. Measures that impede critical research areas can inadvertently diminish national security.

The United States must move decisively to improve significantly the numbers of American-born students who choose science or engineering as a career. This is especially true in the physical sciences, where at present almost half of all Ph.D.s granted in this country are awarded to foreign-born students (Figure 8-2). Yet it is also clear that this situation cannot be changed overnight—it will take many years. While that effort is under way, the country must continue to allow, even encourage, foreign-born students to train here and stay here. U.S. visa policy must be consistent with this goal—a point made forcefully by many U.S. business, academic, and political leaders. At this writing it appears that the government appreciates the need for a careful implementation of visa policy for students, long-term visitors, and permanent residents engaged in U.S. scientific endeavors.

The U.S. scientific community must be fully informed about scientific developments in the rest of the world and must stay fully engaged in international research. Not doing so would have profound ramifications for commerce, health, and defense. To maintain these contacts, foreign visitors must be encouraged to come

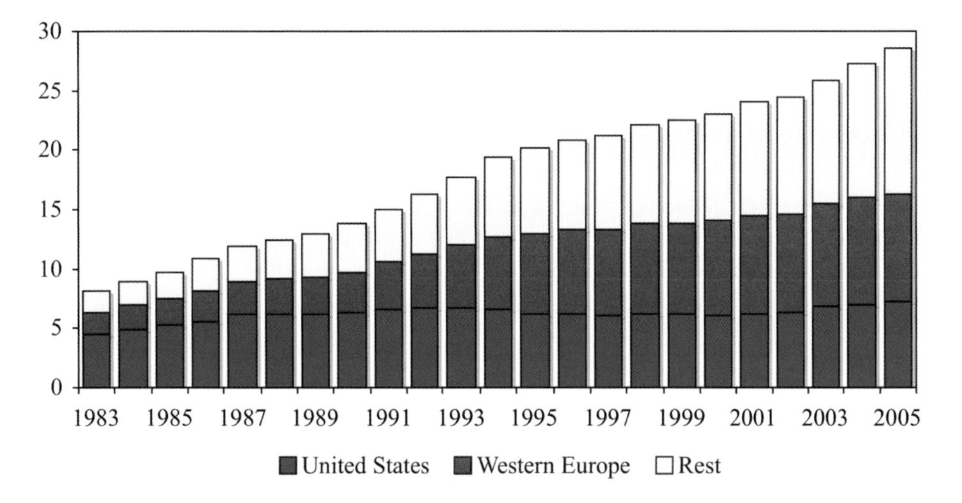

FIGURE 8-4 Number of submissions per year to the world's leading physics journals, *The Physical Review* and *Physical Review Letters*. The U.S. submission rate has remained basically constant since 1987, while the rates for Europe and the Far East have grown rapidly. This trend holds true across all of the subfields of physics, including AMO science. SOURCE: American Physical Society, Publications Office.

to the U.S. to collaborate, attend meetings and conferences, and study. Similarly, U.S. scientists must be able to travel freely abroad for the same purposes. Legitimate security concerns about foreign nationals must not be permitted to weaken the nation's broader, longer-term security goals, which require that we maintain the health and leadership of the U.S. scientific enterprise through international contact.

Another area of great concern to the academic and business community is the so-called "deemed export" restriction.[16] As originally envisioned, this set of rules would have sharply curtailed the ability of foreign students and visitors—that is, nonpermanent residents—to function naturally and effectively in an R&D or educational environment. Given the size of the foreign component of the U.S. research community, these rules as originally stated would have had very serious negative consequences. The committee welcomes the recent and careful consideration by the federal government of how the relevant law can be implemented while mitigating the negative impacts on R&D.

[16]See <http://www.bis.doc.gov/DeemedExports/DeemedExportsFAQs.html#1> for relevant definitions.

PROGRAM CONCLUSIONS ON SUPPORT FOR AMO SCIENCE

The committee draws 10 conclusions based on its analysis:

- **Given the budget and programmatic constraints, the federal agencies questioned in this study have generally managed the research profile of their programs well in response to the opportunities in AMO science. In doing so, the agencies have developed a combination of modalities (large groups; centers and facilities; and expanded single-investigator programs). Much of the funding increase that has taken place at DOE, NIST, and NSF has been to benefit activities at research centers. The overall balance of the modalities for support of the field has led to outstanding scientific payoffs.**
- **The breadth of AMO science and of the agencies that support it is very important to future progress in the field and has been a key factor in its success so far.**
- **Since all of the agencies report that they receive many more proposals of excellent quality than they are able to fund, it is clear that AMO science remains rich with promise for outstanding future progress. AMO science will continue to make exceptional advances in science and in technology for many years to come.**
- **In view of its tremendous importance to the national well-being broadly defined—that is, to our nation's economic strength, health care, defense, education, and domestic security—an enhanced investment program in research and education in physical science is critical, and such a program will improve the country's ability to capture the benefits of AMO science.**
- **Historically, support for basic research has been a vital component of the nation's defense strategy, making the recent decline in funding for basic research at the defense-related agencies particularly troubling.**
- **The extremely rapid increase in technical capabilities and the associated increase in the cost of scientific instrumentation have led to very significant added pressures (over and above the usual CPI inflationary pressures) on research group budgets. In addition, not only has the cost of instrumentation increased, but also the complexity and challenge of the science make investigation much more expensive. This "science inflator" effect means that while it is now possible to imagine research that was unimaginable in the past, finding the resources to pursue that research is becoming increasingly difficult.**
- **In any scientific field where progress is extremely rapid, it is important**

not to lose sight of the essential role played by theoretical research. Programs at the federal agencies that support AMO theory have been and remain of critical importance. NSF plays a critical and leading role in this area, but its support of AMO theoretical physics is insufficient.

- **AMO science is an enabling component of astrophysics and plasma physics but is not adequately supported by the funding agencies charged with responsibility for those areas.**
- **The number of American students choosing physical science as a career is dangerously low. Without remediation, this problem is likely to create an unacceptable "expertise gap" between the United States and other countries.**
- **Scientists and students in the United States benefit greatly from close contact with the scientists and students of other nations. Vital interactions include the training of foreign graduate students, international collaborations, exchange visits, and meetings and conferences. These interactions promote excellent science, improve international understanding, and support the economic, educational, and national security needs of the United States.**

Appendixes

A

AMO 2010 Queries to Federal Funding Agencies

Eight federal agencies (Air Force Office of Scientific Research, Army Research Office, Defense Advanced Research Projects Agency, Office of Naval Research, Department of Energy, National Science Foundation, the National Aeronautics and Space Administration, and the National Institute of Standards and Technology) that fund atomic, molecular, and optical (AMO) research were asked to provide the following information:

1. Data showing the evolution of the size of your agency's AMO program for the past 10 years, in as-spent dollars and 2005 dollars.
2. The number of grantees supported each year for the last 10 years.
3. The number of applicants, and the ratio of successful proposals, each year over the past 10 years.
4. Please provide an estimate of how much turnover there has been in your program over the last few years. In particular, how many new starts have you been able to provide?
5. Please provide a percentage breakdown of how your funds are distributed across the major segments of AMO science today. How much goes to theoretical work, and how much to experiment?
6. What is the average grant size for an award in experimental AMO and in theoretical AMO?
7. Has the number of grants of the same quality level as the ones you have been funding in recent years dropped due to budget shortfalls? How much

would the AMO budget at your agency have to expand to maintain the previous levels of grant numbers?

8. What would be the ideal size of a grant in experimental AMO? In theoretical AMO? How much would it cost to raise all grants to these levels?
9. If you track demographics, what has been the age distribution of the grantees in your program in 2004 and how has that distribution shifted in the last decade?
10. The committee is particularly interested in the percentage participation by women and underrepresented minorities each year for the last 10 years and whether there have been any substantial shifts in the last decade.
11. The number of students supported and Ph.D.s granted each year for the last 10 years.
12. Has your agency begun supporting any special centers or larger group efforts in AMO science? If so, please provide us with a short paragraph or two describing their intellectual thrusts and the size of the award(s).
13. Please describe how the intellectual balance of the support for AMO science has shifted over the last decade. What are the major areas of interest today?
14. Please describe the extent to which your agency supports interdisciplinary activities that include AMO science.

B

Funding

The federal agencies providing support for AMO science do so in a variety of ways: through competitive grants to researchers at universities (NSF or DOE); through grants to peer-reviewed programs at the national laboratories (DOE); through support of in-house research efforts not open to outside competition (largely NIST); or through a mixture of these ways (the defense agencies). In some cases, pass-through funding plays a role. Some agencies rely heavily on peer review, others not at all. These differences are important to keep in mind when making comparisons.

The agencies were asked to provide data showing the evolution in the funding of their AMO programs over the past 10 years (see Table 8-1), to provide information on their average grant sizes for awards in experimental AMO and theoretical AMO science, to estimate the ideal size of a grant in experimental AMO and in theoretical AMO, and to estimate how much it would cost to raise all grants to this size. Finally, they were asked if the number of grants of the same quality level as the ones funded in recent years had dropped due to budget shortfalls, and by how much the AMO budget at their agencies would have to expand to maintain the previous grant levels. The agency responses are given below,[1] and conclusions based on this information are given in Chapter 8.

[1]It was not possible to compare the data this committee received with the data collected for the 1994 FAMOS report, since the research boundaries of the field were defined differently in the two reports. The FAMOS committee appears to have interpreted the field more broadly and, in addition, received more information from the agencies.

DEPARTMENT OF DEFENSE FUNDING AGENCIES

The emphasis at the Air Force Office of Scientific Research (AFOSR), the Army Research Office (ARO), the Defense Advanced Research Projects Agency (DARPA), and the Office of Naval Research (ONR) on engineering applications, often with quick turnaround, has led to significant pressure to downsize funding for basic research (6.1 funds) in favor of funds for applications work (6.2 and 6.3 funds). It was reported to the committee that in many cases 6.1 work requires continuous advocacy to be included in agency programs. The effect is exacerbated by the fact that the overall DOD funding for research has been constant or declining in recent years. These funding shifts have a substantial impact in the university community, which tends to focus on longer-term basic research. Nonetheless, these agencies contribute more than $45 million per year to AMO research activity, roughly $30 million of it from DARPA alone. A significant amount of the DOD funding goes to multidisciplinary research under the Multidisciplinary University Research Initiative (MURI) program. Additional funding to support the purchase of research instrumentation is available to the AMO experimental community through the Defense University Research Instrumentation Program. Through this program, AFOSR, ARO, and ONR awarded more than $43 million in FY2004.

The information from the DOD agencies supplied to us and reported below was mostly anecdotal. Detailed year-by-year trends were not provided.

Air Force Office of Scientific Research

AFOSR funding for its Atomic and Molecular Physics Program has stayed mostly flat over the past decade at about $4 million per year. About 70-75 percent of these funds go to universities. The rest is spent at Air Force research labs such as the one at Wright-Patterson. The program's optical science component is growing and represents about 20 percent of the funding. There is also a separate Optics and Lasers Program that provides about $1.5 million per year for researchers in AMO science. Thus, total funds going to AMO from these two AFOSR programs are about $5.5 million per year. These funds support two MURIs in AMO science at present. There are also AFOSR programs in electro-optics and in nanoelectronics (including work in negative index materials), which are not discussed here.

Average grant sizes in AFOSR core AMO funding are about $125,000 per year for experiment (ranging from about $85,000 to $250,000) and about $100,000 per year for theory. AFOSR staff report that budget pressures have kept grant sizes lower than they should optimally be. Experimental grants could be twice their present size and theoretical ones about 20 percent larger. This would improve research and personnel throughput substantially.

AFOSR is now seeing more proposals from people who used to be funded by ONR and NASA. The quality of proposals is high. Program officers believe that the AFOSR budget for AMO research could easily be doubled before there would be a noticeable change in the quality of work funded.

Army Research Office

ARO core funding in what could be broadly construed as AMO physics has stayed mostly flat over the past decade, at about $2 million per year. This includes some interdisciplinary areas such as photonic band gap materials and imaging science. Basic AMO physics is estimated to be funded at about $1.5 million per year. This also has not changed appreciably over the decade.

Other than core funds, money from special programs, as well as from the Office of the Secretary of Defense, has fluctuated greatly over the decade. Special programs have come and gone. In these programs the actual awards that have gone to AMO-related topics have likewise fluctuated. For example, in the case of MURI awards, there have been anywhere from 1 to 5 active MURIs (at $1 million per year each). Though there have been significant fluctuations, on average about $2 million to $3 million per year of special program monies for AMO has gone to centers of various kinds. At the peak, in 2001, there were three quantum information MURIs focused on AMO topics, two MURIs focused on atom optics, and a number of smaller awards (for equipment, internal centers, etc.). The total funding for special programs amounted to about $5.5 million. Thus, overall AMO physics funding has fluctuated from about $3 million per year at its lowest to about $7 million per year at its peak. For the purposes of this report, the average is taken to be $5 million per year. These funds are spent entirely at universities.

The average grant sizes in ARO core AMO funding are about $140,000 per year for experiment (ranging from about $80,000 to $200,000) and about $80,000 per year for theory (from about $50,000 to $120,000). On the other hand, the MURIs are all fixed in size at $1 million per year. ARO staff believe that grant sizes should all be significantly larger: experimental grants by about 50 percent, and for theory, nearly 100 percent.

There has not been an appreciable drop in the number of grants funded, but the pool of potentially great science has grown, so the AMO budget would have to grow to fund even the very tip of the top proposals. Program officers state that the ARO budget could easily be tripled before the quality of the work funded would slip below current standards.

Defense Advanced Projects Research Agency

DARPA does not maintain a specific program in AMO science. Rather, it involves this work in its characteristic multidisciplinary way as a part of achieving its mission for the DOD (see Appendix D). However, DARPA estimated that funding for AMO-related areas in 2005 totaled over $30 million. This estimate does not include pass-throughs from other agencies, as DARPA typically does not pass funds through but distributes them to other agencies, which act as contracting agents. The complexity and variability of funding patterns at DARPA make it impossible to give an accurate funding history for AMO science from this agency. The funds are predominantly spent at universities, but also by the private sector and by government laboratories.

Because DARPA funds are allocated to achieve a stated mission, it would be misleading to focus on the average size for a research project—there is simply too large a variation in project size and composition for this number to be meaningful. Further, the tendency in DARPA to focus on a particular problem and on multiple disciplines in pursuit of solving that problem often leads to mixing theoretical and experimental tasks within a given research project.

DARPA will continue to be a good, growing source of funding for university AMO researchers, but in a multidisciplinary context, of which AMO science is a very important part. However, there is significant variability in DARPA programs from year to year.

Office of Naval Research

Over the last decade the ONR program in AMO science changed dramatically. The AM component has declined by a factor of 2 and the optical part has been eliminated, leaving about $1.5 million for a core program in AM science only (see also Chapter 8). About two-thirds of the core money goes to university researchers and most of the rest goes to NIST. ONR is unusual in that it does support some overseas performers. ONR also supports four MURIs in AMO science at the present time, as well as a modest Young Investigator Program (each award is for $100,000 per year for 3 years). Total funding of AMO is thus about $5 million per year. Two of the MURIs specialize in the development of new optical frequency standards and optical atomic clocks: one is in quantum control for improved sensing technology and one in new techniques for magnetometry.

Average grant sizes in ONR core AMO funding are about $150,000 per year for experiment (ranging from about $100,000 to $200,000), and about $125,000 per year for theory (ranging from about $100,000 to $150,000). Grant size is sufficient, though program officers believe that the program budget could easily be doubled before there would be a noticeable decline in the quality of work funded.

DEPARTMENT OF ENERGY

The average grant sizes in the DOE's Atomic Molecular and Optical Sciences (AMOS) program are about $137,000 per year for experiment (ranging from about $64,000 to $200,000), and about $104,000 per year for theory (ranging from about $50,000 to $150,000). The ideal size of awards depends on the scope of the project and is best measured in terms of people and equipment available to work on a problem rather than in dollars, because research costs and infrastructure support vary considerably among research institutions. For the typical project that DOE funds at a university, a model grant would provide one month of a faculty member's summer salary and full support for one postdoc and one graduate student plus modest operating support. DOE makes no distinction between experimental and theory efforts for these core elements of a budget plan. For experimental projects DOE anticipates additional equipment and materials costs of about $100,000 per year over the lifetime of a grant, and it encourages the institution to match this investment.

Programs at the national labs are quite different. There, DOE typically supports a number of Ph.D. staff working in concerted fashion on larger-scale integrated programs. Such programs typically have four to seven professional staff and several postdocs.

Funding at DOE for AMOS rose from 1996 to 2005 by about 67 percent in as-spent dollars and by about 34 percent in FY2005 dollars (see Table 8-1). The number of grants supported at DOE increased slightly during that time, and the quality of the work supported has remained high. The concept of "proposal pressure" is not relevant at DOE because of its mission orientation. That is to say, the scientific profile of supported research does not change in response to the level of interest from the researcher community. Nevertheless, within the mission area there is no shortage of excellent proposals, and many strong proposals are not funded.

NATIONAL INSTITUTE OF STANDARDS AND TECHNOLOGY

Funding at NIST for AMOS has risen over the decade by about 55 percent in as-spent dollars and by about 25 percent in FY2005 dollars (see Table 8-1). These funds support a research program that is distributed mostly among six divisions in Gaithersburg, Maryland, and in Boulder, Colorado. NIST does not disburse funds through an unsolicited proposal-based grants program. It does supply some support to universities, about 10 percent of its AMO funds, but this is done in a collaborative way, through both grants and contracts for research and services closely related to its in-house research. At the NIST laboratories, scientists are supported at a level of about $100,000 per year in both experimental and theoretical work, not including overhead or equipment depreciation.

NATIONAL AERONAUTICS AND SPACE ADMINISTRATION

The NASA Laboratory Astrophysics program focuses entirely on atomic and molecular spectroscopy relevant to its overall mission of space exploration. The size of the program is ~$3 million per year, and its focus ranges across wavelengths from x rays to the submillimeter. Grant sizes range from ~$50,000 to ~$300,000. The same grants ranged from $30,000 to $80,000 about a decade ago, so there has been considerable growth over that time. About $2 million per year in the Planetary Atmospheres program is spent on the AMO science relevant to that effort. The Jet Propulsion Laboratory's effort (about $6 million per year) on atomic clocks, atom-wave interferometry, and quantum optics is funded partly by NASA and partly by DARPA. The information from NASA was mostly anecdotal. Detailed year-by-year trends were not provided.

NATIONAL SCIENCE FOUNDATION

The average grant size for an award in the Atomic Molecular and Optical Physics (AMOP) experimental program is about $135,000 per year (ranging from about $100,000 to $300,000). The average grant size for an award in the theory program is about $60,000 per year (ranging from about $35,000 to $90,000). Both programs are in the Physics Division. The committee did not attempt to survey the situation in other parts of NSF (for example, the Chemistry or Materials Research Divisions or in the Computer and Information Science and Engineering Directorate).

Grant sizes for the same level of effort will vary depending on the university, its location, graduate student salaries, indirect cost rates, etc. Averaging over the entire AMOP experimental program, an ideal grant size would be about $170,000 per year. For theory it would be about $100,000 per year. These numbers are calculated based on requested funding levels for proposals currently under review. To achieve this in the experimental program would require about a 25 percent budget increase; the AMO theory program would require about a 50 percent increase. Both assume that the total number of awards does not change.

Budgets in the overall AMOP science program at NSF have increased by about 90 percent in as-spent dollars and about 50 percent in FY2005 dollars over the last decade (see Table 8-1). These funds have come in support of new centers, of individual investigator awards, and to a small extent in special NSF initiatives (for example, in nanoscience and information technology). However, almost all of the increase has gone to the experimental AMOP program. The theoretical side has remained nearly static in spite of the fact that the Physics Division budget is up by nearly 50 percent over this period.

In spite of these increases, the number of grants supported at the top quality levels has declined a bit. The reason for this is that the experimental program has given priority to increasing grant size as opposed to increasing the number of awards. In addition, the program has given attention to providing significant funds for equipment in order to keep funded projects competitive. The budget decreases that took place, particularly in FY2004 and FY2005, impacted this strategy (that is, increasing grant size and providing more funds for equipment) more significantly than they impacted the overall numbers of awards, which have not changed drastically. From FY1998 through FY2002 it was possible to make awards at the level requested, in some cases increasing existing awards by 30-40 percent. Beginning in FY2004 this was no longer feasible. Moreover, while it was possible to provide essentially all the capital equipment requested from FY1998 through FY2002, after that such funding had to be capped at $80,000 per award. Funding for capital equipment is also available through the NSF Major Research Instrumentation program, though the opportunity is somewhat limited because only two requests for funds are allowed per academic institution.

In FY1996 the theory program budget was at about $3.5 million per year. It peaked at $4.7 million per year in FY2001 and has since declined, to about $4.1 million per year in 2005. With advice from the community, the program has elected to keep awards at nearly the existing amount while having to decrease the number of awards. An increase of about 30 percent in the program budget would be needed to return the program to its dollar level in FY2001 and to make up for inflationary losses.

C

Foreign Activity in AMO Science

In this appendix the committee gives a very brief summary of a very large amount of research activity in AMO science taking place in other parts of the world.

FOREIGN-BASED FUNDAMENTAL RESEARCH IN AMO SCIENCE

Are foreign research efforts in AMO science comparable to those in the United States? Unfortunately, the committee lacked the resources to answer this complex question in detail—though it appears that the answer is very clearly yes. In many of the forefront research areas of AMO science—in quantum optics, quantum computing and quantum cryptography, atom cooling and trapping, and fundamentals of quantum mechanics—a significant amount of the leading work is being done outside the United States. Of the 15 Nobel prizes awarded in this area since 1995, 3 of the winners did their work in foreign laboratories and 2 more were trained abroad. AMO science is a highly competitive worldwide activity, with front-rank work taking place in North and South America, Europe, Asia, and the Middle East, in addition to the United States.

While it is certainly true that no single country can yet match the overall U.S. program, the sum of what is taking place in Europe certainly does. The European Union's support for research on a continental scale, combined with the national efforts of its member states, means that Europe is well positioned to compete in research efforts of all kinds. Countries like Germany, with a population of 82 million, and Austria (8 million) have made investments in AMO science that in some cases well surpass U.S. investments even at our most effective and well-funded laboratories (see Box 8-1). The United Kingdom and France, with populations of 60 million each, are not far behind. Japan has committed $15 million to develop a cold atom and quantum information research group at the University of Tokyo. China is forming a 5-year national plan for AMO research, including cold atoms, quantum optics, and quantum information science.

An example of outstanding quality is the Institute for Quantum Optics and Quantum Information, founded in 2003 by the Austrian Academy of Sciences. This institute, located in state-of-the art facilities in Innsbruck and Vienna, is composed of four research groups in experimental and theoretical AMO physics. Following the model of the Max Planck Institutes in Germany, and in particular the Max-Planck Institute for Quantum Optics in Garching (an equally outstanding institute), the institute's goal is to secure a leading role for Austrian science in the fields of quantum optics and quantum information. As a result of this conscious prioritizing and concentrated investment it has succeeded in doing so.

One way to measure of emerging trends in international competition in research is to look at how the number of scientific publications from other countries has grown over the past two decades. Figure 8-4 shows some publication data from two of the world's premier physics journals, *The Physical Review* and *Physical Review Letters.* The relative position of the United States has steadily slipped: in 1990, U.S. scientists accounted for half the submissions, but by 2004 they accounted for only 25 percent of the total. The same trend is visible in AMO science by itself.

To make the discussion more concrete, the committee describes below research facilities in Europe and Asia that are essentially the equal of similarly oriented U.S. facilities. It includes synchrotron light sources, free-electron lasers, and large-scale laser facilities. However, there are also facilities in Europe that have no equivalent in the United States—for example, the heavy-ion storage ring (GSI) in Darmstadt, Germany—that are not discussed. In terms of strategic resources, Europe's proposed global positioning satellite system, Galileo, and Russia's existing one, GLONASS, are fully competitive with the U.S. Global Positioning System (GPS).

FOREIGN-BASED SYNCHROTRON LIGHT AND X-RAY SOURCES

Research with synchrotron light is a rapidly growing, worldwide activity. Today one finds many facilities in Europe, Asia, the Middle East, and South America that are comparable to those in the United States.[1] A significant number of smaller third-generation sources have recently been completed or are currently under

[1]For a comprehensive summary of these laboratories, see <http://www.lightsources.org>.

construction. Here we provide brief descriptions of the most prominent fourth-generation sources slated for completion in the near term.

European X-ray Free Electron Laser Project (Germany)

In February 2003, the German Federal Ministry of Education and Research, together with European partners, approved the XFEL project at DESY. Project R&D will proceed so that a decision to begin construction can be made in the middle of 2006. After a 6-year construction period, the commissioning of the facility could start in 2012. The laser will be based on a roughly 20-GeV superconducting RF linear accelerator. The XFEL, which will operate with ultra-short duration pulses and a brilliance 10^9 times any existing facility, will make it possible to do cutting-edge research in Europe and will guarantee Germany's role as a major location for research and industry.

Fourth Generation Light Source (United Kingdom)

This facility, known as 4GLS, will be a world-leading photon facility that enables internationally outstanding science at Daresbury in the United Kingdom. 4GLS will combine energy recovery linac and free-electron laser technologies to deliver a suite of naturally synchronized, state-of-the-art sources of synchrotron radiation and FEL radiation covering the terahertz to soft x-ray regimes. 4GLS is the leading energy recovery proposal in Europe and the most comprehensive in terms of utilizing combined sources. It is complementary to the European XFEL, to tabletop lasers, and to existing third-generation sources. The project is currently funded through the design and R&D phases.

SPring-8 Compact SASE Source (Japan)

The SPring-8 Compact SASE Source (SCSS) is a high-peak-brilliance, soft XFEL R&D project that aims to operate in the 0.1-nm regime. SCSS will enhance peak brilliance by six orders of magnitude compared to the current third-generation sources in the 3~20 nm range. Like the other FELs discussed here, it produces x-ray light via self-amplified spontaneous emission (SASE).

FOREIGN-BASED LASER LIGHT SOURCES

Laser Megajoule and the High-Energy Multi Petawatt Laser

The French Atomic Energy Center (CEA) is constructing two laser light sources—Laser Megajoule and the High-Energy Multi Petawatt Laser—on the CEA site at Le Barp, Aquitaine. This will be a major research facility comparable to the National Ignition Facility at Lawrence Livermore National Laboratory in the United States, with similar twin goals of national security and fusion energy.

Hundred Terawatt Chirped Pulse Amplified Laser Chain

The Japan Atomic Energy Research Institute (JAERI) has an active research program in atomic physics at high fields. JAERI has constructed a four-stage titanium-doped sapphire chirped-pulse amplification laser system at the Advanced Photon Research Center, Kansai Research Establishment (Kyoto, Japan). The system has achieved a peak power of 0.85 PW in a 33-fs pulse.

FOREIGN GLOBAL POSITIONING SYSTEMS

The GPS has become an indispensable part of the strategic landscape of the U.S. military. Soon such systems also will be a ubiquitous part of public navigational systems worldwide, making possible an enormous variety of useful functions that rely on heretofore inconceivable precision navigation capabilities.

These systems rely heavily on AMO technologies, chiefly in the area of ultraprecise and reliable atomic clocks. Because these clocks are but a small step away from the state of the art (simply because state-of-the-art clocks are not usually ready for long-term launch into space), they give an excellent indication of the technological capabilities of the laboratories that created them. A quick look at the two other global navigation satellite systems in the world shows clearly that their technologies are as good as the technology that is available—or planned—for the U.S. GPS.

GLONASS

GLONASS, Russia's Global Orbiting Navigation Satellite System, is a space-based navigation system comparable to the GPS. It comprises 21 satellites in three orbital planes, with three on-orbit spares. GLONASS provides 100-m positioning accuracy with its deliberately degraded signals and 10- to 20-m accuracy with its military signals. GLONASS has been operational since 1996.

Galileo

Galileo, the European global navigation satellite system, will provide Europe with its own highly accurate, guaranteed global positioning service under civilian control. It is an initiative launched by the European Union and the European Space Agency and will be interoperable with GPS and GLONASS. By offering dual frequencies as standard, however, Galileo will deliver real-time positioning accuracy down to the meter range, which is unprecedented for a publicly available system. It will guarantee availability of the service under all but the most extreme circumstances and will inform users within seconds of a failure of any satellite. This will make it suitable for applications where safety is crucial, such as running trains, guiding cars, and landing aircraft. The fully deployed Galileo system consists of 30 satellites (27 operational and 3 active spares). Once deployed, Galileo will provide a good coverage at latitudes up to 75 degrees north, which corresponds to the northernmost part of Norway and beyond.

D

Intellectual Outlines of Current Research

Each agency was asked to describe the current intellectual portfolio and funding profile for the research it funds today and how that has evolved over the last decade (see also Appendix B). They were also asked to describe what percentage of their funds goes to theoretical and to experimental work and the degree to which their agency supports interdisciplinary activity that includes AMO science. As expected, AMO science is sometimes funded by several parts of a given agency. This is so because on the one hand it is such a broad field and on the other, the support of AMO involved the construction of new facilities. In some cases agencies spend pass-through funds from other agencies on AMO-related activity.

One of the great strengths of the American research enterprise is its breadth. This is very clear in AMO science, with its very rich intellectual prospects, its wide variety of funding sources, and the many modalities available for accomplishing the work. The information provided by the agencies is given below. Conclusions based on it are given in Chapter 8.

DEPARTMENT OF DEFENSE FUNDING AGENCIES

AFOSR, ARO, ONR—located in the Departments of the Air Force, Army, and Navy—and DARPA are all funded by DOD. Their missions, while differing in detail, are broadly directed at protecting U.S. national security. They differ chiefly in the fraction of support they provide to basic, long-range research relative to that for achieving much more immediate goals.

While some of these agencies have programs specifically in AMO science, there is a strong tendency, especially at DARPA, to multidisciplinary work that is difficult to categorize. There is also a lot of emphasis on engineering applications, often with a quick turnaround. To advance the multidisciplinary effort, DOD has created the Multidisciplinary University Research Initiative (MURI) program, which provides a fixed $1 million per year for up to 5 years to each center. Each of the armed services agencies has a MURI component in its program; DARPA does not.

Air Force Office of Scientific Research

AFOSR maintains a relatively stable Atomic and Molecular Physics Program funded from the 6.1 basic research budget. Traditionally this program has not included much optical science, but this component is growing and now represents about 20 percent of the overall effort. There is also a separate Optics and Lasers Program at AFOSR, as well as programs in electro-optics and in nanoelectronics (including work in negative index materials), which is not included in the funding data reported elsewhere in this report.

The philosophy of the AFOSR atomic and molecular physics program is to fund the best science to form a solid research foundation for areas relevant to the Air Force. Recent supported work includes novel methods for ultracooling; precision metrology, including atom interferometry; antihydrogen; optical lattices; optical frequency combs; and electromagnetic induced transparency (slow light). AFOSR supports one MURI in laser diagnostic testing of materials. About 10 percent of AFOSR funds go to theoretical work, but this would be larger with additional funding.

Army Research Office

In rough terms, half of ARO funds for its Atomic and Molecular Physics program and its Optics, Photonics and Imaging program go to atomic and molecular physics and half goes to optical physics. The work is funded from the 6.1 basic research budget. Looking at the data another way, the funds are distributed 80 percent to quantum phenomena (atom optics, quantum optics, degenerate gases) and 20 percent to other fields. About 20 percent of the funding goes to theoretical work.

The recent trend at ARO has been to shift from attempting to cover AMO broadly to supporting more specific research themes. Present areas of interest are atomic and molecular degenerate gases; molecular cooling; optical lattices (for example, for quantum simulations of condensed matter systems); quantum imaging; negative index materials; electromagnetically induced transparency and ultra-broad-band light generation; imaging science broadly; and atom optics. In a typical year, three MURIs are supported. Thus, more than half of the funding is explicitly multidisciplinary.

Defense Advanced Research Projects Agency

DARPA's mission statement sets it apart from other research funding agencies: The mission is to maintain the technological superiority of the U.S. military and prevent technological surprise from harming our national security. DARPA support for any discipline such as AMO physics is constantly shifting. It does this by sponsoring revolutionary, high-payoff research that bridges the gap between fundamental discoveries and their military use. It is by nature highly interdisciplinary (for example, DARPA does not have an identified "AMO program") and does not distinguish between experimental and theoretical work—the research supported will be the research that is necessary to solve the problem at hand. To perform this mission, DARPA manages and directs selected basic and applied research and development projects for DOD and pursues research and technology where risk and payoff are both very high and where success may provide dramatic advances for traditional military roles and missions. The balance of support therefore changes with the ongoing assessment of technology risk and payoff across scientific and technical disciplines. DARPA funds universities, government laboratories, small businesses, and large corporations, utilizing a wide range of contractual vehicles, including grants, but mostly contracts, cooperative agreements, and other transaction agreements.

DARPA has several ongoing programs that support ways to utilize fundamental AMO science in pursuit of DARPA goals. In 2005, these included, among others, the following:

- *Quantum Information Science and Technology.* The program will explore all facets of the research necessary to create a new technology based on quantum information science. Research in this area has the ultimate goal of demonstrating the potentially significant advantages of quantum mechanical effects in communications and computing. Error correction codes, fault-tolerant schemes, and longer decoherence times will address the loss of information. Signal attenuation will be overcome by exploring quantum repeaters. New algorithm techniques and complexity analysis will increase the selection of algorithms, as will a focus on signal processing.
- *Chip-Scale Atomic Clock (CSAC).* The goal is to create ultraminiaturized, low-power atomic time and frequency reference units that will achieve dramatic reductions in size and power consumption, with accuracy specifications similar to present-day rack-mounted devices. The development of

CSACs will enable ultraminiaturized (wristwatch size) and ultra-low-power time and frequency references for high-security, ultra-high-frequency communication and jam-resistant GPS receivers.

- *Precision Inertial Navigation Systems (PINS)* and *Guided BEC Interferometry (gBEC-I)*. The PINS program seeks to use ultracold atom interferometers as an alternative to GPS position updates. Advances in atomic physics over the past two decades have allowed scientists exquisite control over the external quantum states of atoms, including the deliberate production of matter waves from ultracold atoms. This has allowed the development of matter-wave interferometry techniques to measure forces acting on matter, including high-precision atomic accelerometers and gyroscopes. An inertial navigation system that used this technology would have unprecedented low drift rates.
- *Slow Light*. This program is developing functional material systems with the capability to slow and store pulses of light. Perhaps the most dramatic of these slow light pulses to a few meters per second and store and then retrieve a light pulse over several hundred milliseconds. Materials with such capabilities could be used for tunable optical delay lines, optical buffers, high-extinction optical switches, and highly efficient wavelength converters.

Office of Naval Research

The ONR Atomic and Molecular Physics Program focuses entirely on atomic and molecular physics in areas tied to navigation, timekeeping, and sensing applications of relevance to the Navy's mission ("situational awareness"). The optical physics component of the ONR program was eliminated about 3 years ago, in spite of its history of funding some dramatic advances (see Boxes 2-3 and 7-3).[1] ONR supports three MURIs, two in the area of optical frequency standards and atomic clocks, and the third in studies of sub-shot-noise metrology using quantum control techniques. About 33 percent of available funds go to theoretical work.

DEPARTMENT OF ENERGY

DOE's AMOS program in the Office of Basic Energy Sciences (BES) is supported in the context of DOE's overall mission to provide support for energy sciences. It has four thrust areas: Intense Field and Ultrafast X-ray Science, 53 percent of available funds in FY2005; Cooperative, Correlated Phenomena, 28 percent; Ultracold Atoms and Molecules, 12 percent; and Nanoscale Science, 7 percent. These

[1]Chapter 8 includes a discussion about the health of long-term fundamental research at DOD.

are prominent areas of interest in AMO science generally. The AMOS program has seen a significant shift of interest from atomic collision physics and spectroscopy a decade ago to nonperturbative interactions in complex systems and manipulation and control of interactions. The M in AMO science is increasingly prominent in the program (additional information can be found on the Web[2]).

Theory is a significant part of the DOE portfolio. About 38 percent of the current principal investigators (PIs) and co-PIs at universities and DOE's national laboratories are theorists (30 out of 83; however, somewhat less than 38 percent of the funding goes into the theoretical effort). The Chemical Sciences Division at BES has an initiative under way to increase the number of theory projects division-wide. This resulted in three new theory starts in AMOS in FY2005.

Since AMO is an enabling science, multidisciplinary efforts with elements of AMO science play an important role in other DOE programs. For instance, BES draws on AMO science in developing next-generation light sources like the forthcoming LCLS at the Stanford Linear Accelerator Center (SLAC). BES's increasing interest in ultrafast science has led to a number of multidisciplinary efforts with a significant AMO component. Examples include ultrafast beamlines at the Advanced Light Source at the Lawrence Berkeley National Laboratory (LBNL), a new ultrafast x-ray science laboratory at LBNL, and a new ultrafast science center at SLAC. BES programs in nanotechnology, combustion, chemical physics, radiation and photochemistry, heavy element chemistry, separations and analysis, chemical imaging, and interface science all support multidisciplinary efforts that complement the AMOS program and draw on AMO science. Within the AMOS program itself, support is provided for multidisciplinary efforts in nanoscience, photoenergy conversion, imaging at interfaces, and investigation of ultrafast processes in the condensed phase. Outside BES, AMO science has an impact on other DOE multidisciplinary efforts, notably in fusion energy science, environmental remediation, biological science (mainly for imaging), and high energy density science.[3]

NATIONAL INSTITUTE OF STANDARDS AND TECHNOLOGY

The Department of Commerce operates NIST as a part of its mission program in metrology. It maintains three major sites active in AMO physics: the laboratories at its headquarters in Gaithersburg, Maryland; the laboratories at its Boulder, Colorado, facility; and the JILA laboratory on the University of Colorado campus in

---

[2]See <http://www.sc.doe.gov/bes/brochures/BES_CRAs/CRA_13_AMO_Science.pdf>, accessed June 2006.

[3]For more detailed information on the portfolio, such as summaries of projects, locations, and funding for work in specific areas, one can perform key word searches at <http://doe.confex.com/doe/htsearch.cgi>, accessed June 2006.

Boulder. It funds AMO research in six of its divisions: Electron and Optical Physics; Atomic Physics; Optical Technology; Optoelectronics; Time and Frequency; and Quantum Physics.

Support at NIST has increased for laser cooling and trapping, Bose-Einstein condensates, Fermi gas condensation, frequency combs and ultrafast optics, nano-optics, infrared and terahertz radiometry and spectroscopy, optical remote sensing, molecular biophysics, nanofabrication, and nanometrology. Support is down for atomic and molecular collisions (theory and experiment), plasma spectroscopy, x-ray spectroscopy, and atomic and molecular spectroscopy. About 95 percent of NIST funds go to experimental work and 5 percent to theory.

The NIST laboratories have achieved an outstanding record of leadership and productivity in research over the last 10 to 15 years. This has been internationally recognized at the highest level by the award of four Nobel prizes in AMO physics since 1997 to scientists at NIST and the University of Colorado.

In addition to these areas, AMO science plays a very significant role in several other NIST programs in biosystems and health; nanotechnology; homeland security; semiconductor metrology; and quantum information science. An overview of the NIST program can be found in its recent annual report.[4]

NATIONAL AERONAUTICS AND SPACE ADMINISTRATION

The NASA effort in AMO science has changed quite a bit over the last few years with the recent elimination of the program in fundamental physics. For example, the Microgravity program provided very significant support for front-rank research in both the AMO and condensed matter physics communities, so its termination is a serious blow to the NASA program. In addition, the gravity program contains only trace amounts of AMO science.

Today, the Laboratory Astrophysics program exists primarily to supply spectroscopic data in support of NASA missions. Hence it concentrates on the atoms and molecules that exist in a space environment and that emit and absorb at wavelengths accessible to NASA missions. Similarly, about 20 percent of the various Planetary Sciences programs could be called "applied atomic physics" because they focus on such issues as molecular and atomic spectra at extremes of temperature and pressure. However, there is concern that the level of support for laboratory astrophysics may not be sufficient for it to play its essential scientific role in the programs of a number of prominent astronomy facilities (see Box D-1).

---

[4]For more information see <http://www.physics.nist.gov/TechAct.2004/NIST-PhysLab2004TechAct.pdf>.

BOX D-1
Atomic and Molecular Astrophysics Research Support

As noted above and in Chapter 2, atomic and molecular studies of a variety of collision processes and spectroscopy are of critical importance to astronomy. Although one to two decades ago support for research in these areas made up a large fraction of the AMO physics funding portfolio at NSF and DOE, these agencies currently fund very little of this research. This decline in support for atomic and molecular astrophysics ("laboratory astrophysics") at these agencies is due to the competition for AMO physics support from the many other forefront areas described in this report.

However, some laboratory astrophysics continues to be funded at NASA (~\$4 million) in support of NASA missions. But considering the current and future needs of the astrophysics community, especially as newer, very powerful ground-based and space-based instruments with greater photon sensitivity and higher spectroscopic resolution are developed, the field of atomic and molecular astrophysics may not be able to support the use of these new instruments to their fullest capacities. There is concern that the field of laboratory astrophysics is not currently training enough graduate students and postdoctoral fellows to maintain research expertise as senior scientists in this field retire. In view of the very large federal investment in such instruments and facilities as the Chandra X-ray Space Telescope; the Atacama Large Millimeter Array (ALMA), under construction; and the James Webb Space Telescope (scheduled for launch in 2013), this issue must be examined carefully, across the agencies, in the context of other priorities in AMO science and in astronomy.

Work on atomic clocks is continuing at the Jet Propulsion Laboratory (JPL), but it is now of a much different character. The goal is no longer ultrahigh stability but rather the construction of small stable clocks for space missions. There is also ongoing work on atom-wave interferometers on a chip, again for space missions. JPL is also vigorously pursuing work in quantum optics.

NATIONAL SCIENCE FOUNDATION

The NSF's Atomic, Molecular, Optical, and Plasma Physics (AMOP) program encompasses four areas: Precision Measurements, 22 percent of available funds in FY2005; Atomic and Molecular Dynamics (mostly support of BECs and related subjects), 42 percent; Atomic and Molecular Structure, 12 percent; and Optical Physics, 24 percent. Support in the first three areas includes activities in quantum control, cooling and trapping of atoms and ions, low-temperature collision dynamics, the collective behavior of atoms in weakly interacting gases (BECs and dilute Fermi degenerate systems), precision measurements of fundamental constants, and the effects of electron correlation on structure and dynamics. In optical physics, support is provided in areas such as nonlinear response of isolated atoms to intense, ultrashort electromagnetic fields, the atom-cavity interaction at high fields, and quantum properties of the electromagnetic field. The AMO theory program covers the same broad areas.

NSF, unlike the other agencies discussed, is not a mission-oriented agency. Rather, it shapes its scientific portfolio based mostly on the scientific interest and merit of the proposals submitted by the community. Observation of Bose-Einstein condensation (BEC) in 1995 led to growth in that area. Thus the more mature areas of AMO research in individual-particle collisions, such as electron-atom scattering and ion scattering, have become less active than research in cold collisions and phenomena related to BEC. This trend continues in that BEC per se is no longer of special interest. Research is moving onward to the study of collective effects in quantum fluids, for example, at the Physics Frontier Center for Ultracold Atoms.

Advances in laser technology also drove increased support in the area of quantum control, including the large Physics Frontiers Centers award to the Frontiers in Optical Coherent and Ultrafast Science (FOCUS) program. Finally, the impetus provided by quantum information science has driven an increase in the number and quality of proposals in optical physics, a trend that has become increasingly manifest in the past 3 years. It also led to enhanced funding through the NSF-wide programs Information Technology Research (ITR) and Nanoscale Science and Engineering (NSE). Support for research in atomic and molecular structure, which is primarily spectroscopy, also dropped over the last decade. The support for research in precision measurements has remained essentially constant. The distribution of funding between experiment and theory varies from year to year, with theory hovering at between 15 and 20 percent of experiment.

As indicated, AMO science is included in NSF ITR and NSE priority areas, though the funding impact has not been large. AMO-related science is also funded through the Divisions of Chemistry and Materials Research in the Directorate for Mathematical and Physical Sciences and to some extent also in the Directorate for Computer and Information Science and Engineering and the Directorate for Engineering.

E

People

The agencies were asked to provide information for each of the last 10 years on the number of grantees supported, the number of applicants for that support, and the success rates. They were also asked to comment on turnover, age distribution, and participation by women and minorities over the last decade and whether there have been any noticeable shifts. Finally, they were asked about the number of students supported (graduate and undergraduate), Ph.D.s granted, and postdocs supported each year over the last 10 years. These data are compiled in Table 8-2 and in Figures 8-2 and 8-3, and the committee's conclusions appear in Chapter 8.

Federal law prohibits collecting some of this information except via voluntary self-reporting, so the discussion below is based on qualitative impressions from the program officers.

DEPARTMENT OF DEFENSE FUNDING AGENCIES (AFOSR, ARO, DARPA, ONR)

DOD programs operate with several differing proposal review systems, including peer review. For example, the MURI program is reviewed like NSF programs. Other awards are based on their suitability for armed services programs. Program officers have large discretionary powers in making the funding decisions in most cases.

Air Force Office of Scientific Research

Average grant size at the AFOSR is about $125,000 per year, so about 20 university PIs are in the program at any one time. Other AFOSR funds support AMO research done at Air Force labs. University funding applications proceed along both informal and formal lines, often beginning with a telephone call or an e-mail. Most ideas are turned away at this stage. But if the idea is interesting, a white paper is solicited or a proposal is requested. If a proposal is requested funding will be supplied if at all possible. Young people are encouraged. Program turnover is about 5 percent per year.

AFOSR does not track demographics in detail, but estimates are that roughly 10 percent of grantees are women and about 7 percent are minorities. The agency does send some funding to Historically Black Colleges and Universities (HBCUs). Roughly speaking, each PI has an associated graduate student, so about 20 are supported in the program. Roughly four doctoral degrees are granted each year.

Army Research Office

A good rule of thumb at the ARO is that there is roughly one grantee per $115,000 of funding (with a range of $80,000 to $200,000). Thus on average, ARO is supporting somewhere between 20 and 60 senior investigators depending on the year. The number of awards hovers around 30. Almost all applications start with a phone call or e-mail. Perhaps 90 percent are not encouraged any further. Of the applicants who submit white papers, perhaps 30 to 50 percent are encouraged to submit a proposal. Of submitted proposals, perhaps 30-50 percent are funded.

Turnover occurs as a result of deliberate changes in direction or a deliberate desire to change the mix of people. Over the last 2 years 50 percent of the atomic and molecular part of the core program has turned over. The Special Programs typically last 3-5 years and are not renewable. Thus they automatically turn over. The average number of new starts per year is perhaps 5 (out of the ~30 awards in place on average). ARO is constantly introducing new young investigators to the program while a seasoned crew ages.

Like its sister agencies, the ARO does not track demographics in detail. Women have averaged about 10 percent of the program over the years. The number of underrepresented minority PIs is probably less than 5 percent but with only about 30 awards the statistics are too poor to be more than simply indicative of very low participation.

Roughly speaking, each $100,000 award corresponds to one graduate student. With an average budget of about $5 million per year, this is approximately 50 graduate students. ARO does not track the number of Ph.D.s awarded but guesses it would be about 10 per year. ARO supports about 30 postdocs each year.

Defense Advanced Research Projects Agency

For the most part, DARPA does not award grants, but it does award contracts. Most of these are awarded and managed through AFOSR, ARO, and ONR.

Office of Naval Research

The average grant size at ONR is about $150,000 per year, so about 30 university PIs are in the program at any one time. Application procedures are essentially the same as at the other DOD agencies.

ONR does not track demographics in detail, and outreach activity is not measured, as it is not a part of the ONR mission. ONR does support some students and postdocs as a normal part of conducting its mission, but it is difficult to estimate reliably how many.

DEPARTMENT OF ENERGY

The number of awards at universities fluctuates from year to year, but the average has remained relatively constant over this period. There are multiple PIs on some grants, and a number of PIs are supported by programs at DOE national laboratories. Currently the total number of PIs and co-PIs in the program is 83, including experiment and theory. About 38 percent of PIs are in the theoretical program, but somewhat less than 38 percent of the funds goes to theoretical work.

A typical DOE grant provides full or partial support for a postdoc and one or two graduate students, so at any time there are roughly 150 graduate students and postdocs working with full or partial support from DOE—that is, about 100 graduate students and 50 postdocs. There are some undergraduates working in DOE's AMOS program, but it is very difficult to estimate the number. It is not large.

Since DOE is mission-oriented, it solicits and considers only proposals in certain areas, so that general notion of proposal pressure from the AMO community is not relevant. It does receive many inquiries about whether DOE's AMOS program would consider research in certain areas, and many of these are discouraged, though the numbers are not tracked. In the past 2 years, about 20 percent of the encouraged proposals were successful.

The number of grants turning over fluctuates from year to year. In FY2005 there were five new starts and in FY2004 there were two, and in that 2-year period seven grants ended. For the years 2002 through 2004, 83 percent of the grants coming up for renewal (typically after 3 years of funding) were renewed (30 out of 36). In that same period, 34 percent of the new proposals considered were successful (15 out of 44).

Like its sister agencies, DOE does not track demographics, so no quantitative information was provided on trends for women or minorities. But the qualitative belief is that these numbers are low, consistent with the experience of the other agencies.

NATIONAL INSTITUTE OF STANDARDS AND TECHNOLOGY

The six NIST divisions contributing to AMO science in Gaithersburg and Boulder employ among them about 200 scientists, 20 postdocs, 30 graduate students, and 20 undergraduates. In addition, there are about 100 guest scientists, mostly from abroad, who collaborate with the permanent NIST staff.

NATIONAL AERONAUTICS AND SPACE ADMINISTRATION

In recent years the NASA Laboratory Astrophysics program has supported about 30 senior investigators and a few postdocs and graduate students. The Planetary Atmospheres program supports one-third to one-half that number. The JPL program, which is directed entirely in-house, supports about four graduate students but no postdocs. A problem that NASA foresees is that not enough young people are coming into the area of applied astrophysics. Most of the work is done at centers or at private research institutes, which tend not to focus on training students.

NATIONAL SCIENCE FOUNDATION

Between FY1997 and FY2005, NSF's AMOP experimental program supported an average of 102 senior investigators each year (including PIs and co-PIs) under roughly 86 active awards. However, the average number of senior personnel supported from FY1997 to FY1999 was about 15 percent higher than the average from FY2003 to FY2005 (109 vs. 93). Similarly, the number of postdocs supported declined by about 23 percent over the period, from 51 to 39. However, the number of Ph.D. students supported increased by 18 percent, from 132 to 156. Over the same period the AMOP theory program supported an average of 52 senior investigators, 19 postdocs, and 41 graduate students under about 50 active awards. These numbers have remained essentially constant over the decade. Neither program tracks the numbers of Ph.D.s awarded in any year. An estimate is that about 25-30 doctoral degrees are granted each year in experiment and about 8 in theory.

Between FY2000 and FY2005, the AMOP experimental program supported an average of 97 undergraduate students each year through Research Experiences for Undergraduates (REU) supplements and through direct awards. In the same period the AMOP theory program supported an average of 16 undergraduate students each year.

Beginning in FY1997, the AMOP experimental program began giving preference to increasing the size of awards as opposed to increasing the number of awards or the number of investigators supported. This practice has since been maintained. On the other hand, because funding for theory has recently been static or declining, the program has chosen to try and maintain a roughly constant award size. The AMOP experimental and theoretical programs averaged 41 and 46 percent success rates, respectively, over the reporting period.

Both AMOP programs place high value on initiating new grant activities, particularly those of junior investigators. Of new awards made in the experimental program between FY2000 and FY2005, 34 were to scientists within 10 years of their Ph.D. The numbers vary, from four new starts in FY2001 and FY2005 to eight new starts in FY2002. The much smaller AMOP theory program reports a few new starts each year. One reference point is illustrative: Of the 99 scientists listed as PIs on experimental awards in FY1996, 53 were listed as such in FY2005.

Like its sister agencies, the NSF does not track demographics in detail. Like the other agencies, the qualitative response is that the AMOP experimental program is weak with regard to participation by women and underrepresented minorities. Of the 102 senior investigators reported above, eight were women and two were underrepresented minorities. The number of women supported increased over the decade, from four to eight. The number of underrepresented minorities remained the same.

The AMOP theory program has similar experience with women as senior investigators. Of the 52 senior investigators mentioned above, 5, on average, have been women. This number has increased slightly in recent years. One underrepresented minority was funded from FY1999 through FY2002. During the remaining years none were reported. These numbers reflect only self-identified individuals.

F

New Research Modalities

New modalities for doing science arise as a means of doing science more effectively. The best ones arise naturally, following the needs of the science. One particularly important reason for collaboration is the increasingly interdisciplinary nature of the work being done throughout science. Another reason is the availability of large-scale facilities (synchrotron light sources, high-intensity lasers, specialized laboratories), which provide unique access to specialized instrumentation and expertise. Yet another is the value of assembling a critical mass of people to work on closely related topics.

DEPARTMENT OF DEFENSE

The Multidisciplinary University Research Initiative (MURI) concept is unique to the DOD funding agencies (AFOSR, ARO, DARPA, and ONR). These 5-year awards of $1 million per year are intended to advance the necessary research in a university environment. They are in wide use.

Air Force Office of Scientific Research

As described in Appendix D, the AFOSR supports one MURI effort in laser diagnostics. It supports no other centers.

Army Research Office

A substantial fraction (60-70 percent) of ARO funding goes into centers whose topics of study have changed over the decade. Recently, MURI centers have been studying quantum imaging (employing entanglement to perform nonclassical imaging, including subwavelength resolution, ghost imaging, quantum radar, pixel entanglement, and so on); atom optics (quantum degenerate gases such as Bose condensates, atom lasers, and the like, and guiding them in free space and on chips, performing interferometry, and so on); and quantum information and computing (exploiting quantum entanglement in ion traps, optical lattices, molecules, and so on for making qubits, teleporting information, transferring coherence between entities as in from photons to degenerate gases and back, cavity QED implementations, etc.). Quantum information and computing has seen numerous MURIs come and go. The Army also supports a small in-house optics center (not a MURI) at West Point.

Defense Advanced Research Projects Agency

DARPA is unique in DOD in that it does not maintain an infrastructure of laboratories or research facilities. This allows it to minimize institutional interests that would otherwise distract it from its search for new research areas and world-class performers. DARPA does not necessarily seek to advance progress in established disciplines, but instead will bring together teams from diverse institutions and disciplines to solve a particular problem. In the past, this was sometimes done through the establishment and funding of interdisciplinary laboratories. More recently, DARPA has been funding interdisciplinary teams of researchers from multiple research institutions without establishing a fixed infrastructure. Currently, DARPA does not operate or fund any centers in AMO science, though it funds major AMO-related collaborations.

Office of Naval Research

The ONR presently supports three MURI awards at $1 million per year each. Two are in optical frequency standards and atomic clocks, while the third studies sub-shot-noise measurement using quantum control.

DEPARTMENT OF ENERGY

The AMO science program supports five large group efforts. Four are at DOE national laboratories and one is at Kansas State University. The levels quoted are the FY2005 allocations. Yet another—the Photon Ultrafast Laser Science and Engineering (PULSE) program at Stanford—is supported from the BES Materials Science Program.

- *J.R. Macdonald Laboratory Program for Atomic, Molecular, and Optical Physics at Kansas State ($2.5 million per year).* This laboratory combines theory and experiment to investigate dynamical processes involving ions, atoms, molecules, surfaces, and nanostructures exposed to short, intense bursts of electromagnetic radiation. Current efforts are focused on time-resolved dynamics of heavy-particle motion in single molecules and molecular ions; coherent excitation and control in multilevel systems; interaction of intense short-pulse laser radiation and ions with surfaces and nanostructures; attosecond science (in particular using real-time probes of the electronic wave function); and collisions with highly charged ions.
- *Multiparticle processes and interfacial interactions in nanoscale systems built from nanocrystal quantum dots at Los Alamos ($0.8 million per year beginning in 2002).* This research is aimed at controlling the functionalities of nanomaterials. It requires a comprehensive physical understanding at different levels, ranging from individual nanoscale building blocks to the complex interactions in the nanostructures built from them. This project concentrates on electronic properties of semiconductor quantum-confined nanocrystals and the electronic and photonic interactions of assemblies of them. The group studies multiparticle processes in individual nanocrystals and interfacial interactions. The ability to understand and control both multiparticle processes and interfacial interactions could lead to such new technologies as solid-state optical amplifiers and lasers, nonlinear optical switches, and electrically pumped, tunable light emitters.
- *Atomic and Molecular Physics Group at Oak Ridge National Laboratory (ORNL) ($1.8 million per year).* The goal is the understanding and control of interactions and states of atomic-scale matter. The objective is to develop a detailed understanding of the interactions of multicharged ions, charged and neutral molecules, and atoms with electrons, atoms, ions, surfaces, and solids. Toward this end, a robust experimental program is carried out at the ORNL Multicharged Ion Research Facility and as needed at other facilities. Closely coordinated theoretical activities support this work, as well as lead investigations in complementary research. Specific focus areas for the program are broadly classified as particle-surface interactions, atomic processes in plasmas, and manipulation and control of atoms, molecules, and clusters.
- *Atomic, Molecular and Optical Sciences Group at Lawrence Berkeley National Laboratory ($1.365 million per year).* This program is aimed at understand-

ing the structure and dynamics of atoms and molecules using photons and electrons as probes. The current emphasis is in three major areas with important connections and overlap: inner-shell photoionization and multiple ionization of atoms and small molecules; low-energy electron impact and dissociative electron attachment of molecules; and time-resolved studies of atomic processes using a combination of femtosecond x rays and femtosecond laser pulses. The goal of the ultrafast science effort is to probe fundamental atomic and molecular processes involving femtosecond (and ultimately attosecond) x rays interacting with atoms and molecules in the presence of laser fields and to shed light on electron correlations within these systems.

- *Atomic, Molecular, and Optical Physics Group at Argonne National Laboratory ($1.215 million per year).* The central goal is to establish a quantitative understanding of x-ray interactions with free atoms and molecules. With the advent of hard x-ray free electron lasers, exploration of nonlinear and strong-field phenomena in the hard x-ray regime becomes possible. Techniques for microfocusing x rays are being developed to help understand the behavior of atoms and molecules in strong optical fields. These studies of atoms and molecules in strong optical fields will be relevant for pump-probe experiments at next-generation sources. Foundational to these experiments is the detailed understanding of x-ray photoionization. The group has focused on the limitations of our current theoretical understandings.
- *The Photon Ultrafast Laser Science and Engineering Center at SLAC.* The Stanford PULSE Center conducts interdisciplinary research in ultrafast science. PULSE has a major AMO component, although in 2005 it was funded from the Materials Science program in DOE. A major AMO focus of this center is research at the LCLS x-ray free-electron laser, specifically on the interaction of atoms and molecules with high-field and ultrafast short-wavelength radiation and on the control of quantum dynamics in atoms and molecules at attosecond to femtosecond timescales.

NATIONAL INSTITUTE OF STANDARDS AND TECHNOLOGY

NIST supports six divisions at its Gaithersburg and Boulder sites in which AMO science plays a lead role. Some of the funding for these laboratories comes from other federal sources (for example, DOD and NASA). Relatively new areas of AMO competence, with total funding levels, are these:

- Quantum information/quantum computing/quantum communication ($9 million per year).

- The Center for Nanoscale Science and Technology (CNST) user's facility, located within the new NIST Advanced Measurement Laboratory (500,000 ft^2, featuring a 90,000 ft^2 clean room facility, built at a cost of $200 million). The annual budget for CNST is about $6 million.
- Molecular measurement and manipulation (about $1 million per year, aimed at bioscience).
- Low-temperature quantum coherence, including laser cooling and trapping, Bose-Einstein and Fermi condensation, atomic fountain clocks, trapped ion optical clocks (about $7 million per year).

NATIONAL SCIENCE FOUNDATION CENTERS IN AMO SCIENCE

NSF's Mathematical and Physical Sciences Division supports four centers in AMO science. The funding levels quoted below are the FY2005 allocations.

- *Center for Ultracold Atoms (CUA) ($1.5 million per year beginning in FY2000).* CUA is funded through the AMOP program in the Physics Division and the Condensed Matter Physics program in the Division of Materials Research. In FY2005, it was transferred to the Physics Frontiers Centers program for award management and future competition. CUA brings together a community of scientists from the Massachusetts Institute of Technology and Harvard University to pursue research in the new fields. The core research program consists of four collaborative experimental projects whose goals are to provide new sources of ultracold atoms and quantum gases and new types of atom-wave devices. These projects will enable new research on topics such as quantum fluids, atom/photon optics, coherence, spectroscopy, ultracold collisions, and quantum devices. In addition, the CUA has a theoretical program centered on quantum optics, many-body physics, wave physics, and atomic structure and interactions.
- *Frontiers in Optical Coherent and Ultrafast Science (Physics Frontier Center, $15 million over 5 years beginning in 2001).* The FOCUS mission is to provide national leadership in the areas of coherent control, ultrafast physics, and high-field physics. FOCUS will extend the frontiers of the discipline: the production, control, and utilization of subpicosecond and, eventually, subfemtosecond pulses; coherent manipulation of molecular bonds and intramolecular dynamics; physics of ultrahigh laser fields (luminosity $> 10^{20} W/cm^2$); and control of entanglement in ultracold atoms and ions. The coherent field strengths under direct control will span 18 orders of magnitude, from ultrarelativistic, laser-driven plasmas (teravolts per centimeter) to control fields in cooled ion traps (millivolts per centimeter).

Laser-driven particle energies will range from GeV to neV. Much of the coherent control physics developed in one area is applicable to other areas.

- *JILA (jointly supported with NIST; $3.2 million from NSF in 2005).* Although the majority of its funding comes from NIST, NSF's AMOP program provides an amount of funding comparable to that for any of its other large centers. NSF will begin to treat JILA in the same category as FOCUS and CUA starting in the 2006 renewal cycle.
- *Institute for Theoretical Atomic, Molecular and Optical Physics (ITAMP; Harvard-Smithsonian Center for Astrophysics and the Harvard University Physics Department; $0.65 million per year, begun in 1988).* Though not funded as a "center" by NSF, it functions as one as far as the theoretical AMO community is concerned. It has very active postdoctoral and visitor programs, a constantly changing menu of active theoretical topics, a lively series of workshops and an excellent computation center. It prospers significantly from its close ties to the Harvard-Smithsonian Center for Astrophysics.
- *Engineering Research Center for Extreme Ultraviolet Science and Technology (Colorado State University, University of Colorado at Boulder, University of California at Berkeley, and Lawrence Berkeley National Laboratory).* Funded by the NSF's Directorate for Engineering, the initial award ($17 million over 5 years beginning in October 2003) supports the first 5 years of a 10-year cooperative agreement. The goal of the center is to confront a variety of challenging scientific and industrial problems by using short-wavelength light in the extreme ultraviolet (EUV) range of the electromagnetic spectrum. The researchers are exploring the interface of physics, electrical engineering, chemistry, and biology using high-energy, extremely short-wavelength, coherent EUV light.

U.S. SYNCHROTRON LIGHT AND X-RAY SOURCES

Over the past two decades, AMO science has played an important role in the development of very intense synchrotron light sources[1] that have been indispensable not only to AMO researchers but also to workers in materials science, condensed matter physics, and biology. In the United States, the forefront laboratories are these:

- Advanced Photon Source at Argonne National Laboratory, Illinois (third-generation x-ray source).[2]

---

[1]See <http://www.lightsources.org> for information on the many facilities operating worldwide.

[2]First-generation sources are high-energy physics machines with parasitic operation for synchrotron radiation. Second-generation sources are dedicated machines with bending magnet beam lines.

- Advanced Light Source (ALS) at Lawrence Berkeley National Laboratory, California (third-generation soft x-ray source).
- Cornell High Energy Synchrotron Source at Cornell University, Ithaca, New York (third-generation x-ray source).
- National Synchrotron Light Source (NSLS) at Brookhaven, New York (second-generation VUV and x-ray source).
- Stanford Synchrotron Radiation Laboratory's Stanford Positron Electron Accelerating Ring at Menlo Park, California (third-generation x-ray source).

Other smaller facilities in the United States are these:

- Center for Advanced Microstructures and Devices, Baton Rouge, Louisiana (second-generation soft x-ray source).
- Duke Free Electron Laser Laboratory, Durham, North Carolina (fourth-generation infrared source).
- Jefferson Laboratory Free Electron Laser, Newport News, Virginia (fourth-generation infrared source).
- Synchrotron Radiation Center, Madison, Wisconsin (third-generation VUV source).
- Synchrotron Ultraviolet Radiation Facilty, NIST, Gaithersburg, Maryland (second-generation VUV source).
- UCSB Center for Terahertz Science and Technology, Santa Barbara, California (fourth-generation far-infrared source).
- W.M. Keck Free Electron Laser Center at Vanderbilt University, Nashville, Tennessee (fourth-generation mid-infrared source).

The fourth generation of x-ray light sources will not be synchrotrons at all but will be built around the concept of the free-electron laser[3] using linear accelerators. Worldwide there is very substantial R&D work in progress in this area (see Appendix C).

Linac Coherent Light Source at Stanford

The LCLS, a $379 million facility, will be the world's first x-ray free-electron laser facility for science when it becomes operational in 2009. Construction started

Third-generation sources are machines designed to make maximal use of insertion devices called "undulators" and "wigglers." All the recent facilities, beginning approximately with ALS in Berkeley, are built this way. Fourth-generation sources are free-electron lasers.

[3]See <http://sbfel3.ucsb.edu/www/vl_fel.html> for links to these activities.

in FY2005. Pulses of x-ray laser light from LCLS will be many orders of magnitude brighter and several orders of magnitude shorter than what can be produced by any other x-ray source available now or in the near future. These characteristics will enable frontier new science in areas such as discovering and probing new states of matter, understanding and following chemical reactions and biological processes in real time, imaging chemical and structural properties of materials on the nanoscale, and imaging noncrystalline biological materials at atomic resolution. The LCLS project is funded by the DOE/BES and is a collaboration of six national laboratories and universities.

LARGE-SCALE LASER FACILITIES

There are also a number of large-scale laser facilities with strong links to AMO, built here and abroad (see Appendix C).

- *OMEGA at the Rochester Laboratory for Laser Energetics.* This facility is the highest energy laser working in the United States. DOE operates OMEGA primarily for work related to laser fusion. University researchers are also eligible to apply for shots on OMEGA.
- *JANUSP at the Lawrence Livermore National Laboratory.* This facility is used for high-field laser-atom and laser-plasma physics by scientists at universities as well as the national laboratories.
- *National Ignition Facility (NIF).* When completed, NIF at the Lawrence Livermore National Laboratory will be the nation's most powerful laser, supporting DOE's National Nuclear Security Administration defense program's mission to verify the safety and reliability of U.S. nuclear weapons. NIF will also study inertial confinement fusion for energy.